D0884943

NUMBER THEORY

GEORGE E. ANDREWS

Evan Pugh Professor of Mathematics
Pennsylvania State University

DOVER PUBLICATIONS, INC.
NEW YORK

Bibliographical Note

This Dover edition, first published in 1994, is an unabridged and corrected republication of the work first published by W.B. Saunders Company, Philadelphia, in 1971.

Library of Congress Cataloging-in-Publication Data

Andrews, George E., 1938–
Number theory / George E. Andrews. — 1st Dover ed.
p. cm.
Originally published: Philadelphia : Saunders, 1971.
ISBN-13: 978-0-486-68252-5 (pbk.)
ISBN-10: 0-486-68252-8 (pbk.)
1. Number theory. I. Title.
QA241.A5 1994
512′.7—dc20 94-5243
CIP

Manufactured in the United States by LSC Communications
68252817 2020
www.doverpublications.com

To Joy and Amy

NOTE ON THE COVER

The cover illustration is an artist's representation of the so-called golden rectangle. If the largest square in a golden rectangle is cut away, then the figure remaining will also be a golden rectangle. Such rectangles are characterized by a length to width ratio of $(1 + \sqrt{5})/2$, the golden ratio. The ancient Egyptians may have used this ratio in the construction of pyramids. The ratio recurs often in number theory; for example,

$$\lim_{n \to \infty} \frac{D_2(n)}{D_2'(n)} = \lim_{n \to \infty} \frac{F_{n+1}}{F_n} = \frac{1 + \sqrt{5}}{2},$$

where $D_2(n)$ and $D_2'(n)$ are the partition functions occurring in the Rogers-Ramanujan identities, and F_n is the nth Fibonacci number.

PREFACE

Most mathematics majors first encounter number theory in courses on abstract algebra, for which number theory provides numerous examples of algebraic systems, such as finite groups, rings, and fields. The instructor of undergraduate number theory thus faces a predicament. He must interest advanced mathematics students, who have previously studied congruences and the fundamental theorem of arithmetic, as well as other students, mostly from education and liberal arts, who usually need a careful exposition of these basic topics.

To interest a class of students whose backgrounds are so divergent, this text offers a combinatorial approach to elementary number theory. The rationale for this point of view is perhaps best summarized by Herbert Ryser in *Combinatorial Mathematics:* "... combinatorics and number theory are sister disciplines. They share a certain intersection of common knowledge and each genuinely enriches the other."* In studying number theory from a combinatorial perspective, mathematics majors are spared repetition and provided with new insights, while other students benefit from the consequent simplicity of the proofs for many theorems (the proofs of Theorems 1–3, 3–4, 3–5, 6–1, 6–2, 6–3, 7–6, 8–4, and 9–4 rely mainly on simple combinatorial reasoning). Number theory and combinatorics are combined in Chapters 10 through 15 to aid in the discovery and proof of theorems.

Two aspects of the text require preliminary discussion. First, Section 5 of Chapter 3 is critical to the whole work. This section illustrates both the value of numerical examples in number theory and the role of computers in obtaining such examples. The accompanying exercises provide opportunities for constructing numerical tables, with or without a computer. Subsequent chapters may then be introduced to advantage by allowing students to report on conjectures they derive from relevant numerical tables. When students are thus actively involved, theorems will seem natural and well motivated.

*Ryser, Herbert J., *Combinatorial Mathematics* (Carus Monograph No. 14). Mathematical Association of America, 1963.

Second, in Chapters 12, 13, and 14, the student will encounter partitions, a topic in additive number theory. Too often, one obtains from number theory texts the impression that each topic has been thoroughly developed. The problems offered in such texts are either solved or unsolvable; at best, the student is invited to work a few peripheral problems. In this book, Chapter 12 attempts to communicate the excitement of the mathematical chase by devising a procedure for forming conjectures in partition theory. The exercises at the end of Chapter 12 provide the student with a number of opportunities for discovering theorems himself. Chapters 13 and 14 develop techniques in the application of generating functions to partition theory so that the student can prove some of the conjectures he made in Chapter 12. In presenting Chapter 14, the instructor should assign the exercises at the end prior to beginning lectures, in order to avoid the unmotivated presentation of complicated manipulations of series and products; through this procedure, the student is led to appreciate the relation between the exercises and the steps in the proof of the Rogers-Ramanujan identities and of Schur's Theorem.

Many people have aided me in preparing this book. I express particular thanks to the students in my class of Spring Term, 1970, at Penn State, who were taught from the completed text and who generously offered valuable suggestions. Professors H. L. Alder, G. L. Alexanderson, and G. Piranian read the entire manuscript and made many useful contributions. Carlos Puig and George Fleming of W. B. Saunders have skillfully guided the process of publication.

Finally I pay tribute to my wife, Joy, who has been the most important helper in the creation of this book; at each stage her encouragement, intelligence, and energy have added significant value. She has been immensely creative both in writing expository material and in facilitating the communication of ideas to students. Without her aid, a mass of scribbled lecture notes would still be just that.

For certain classes where the instructor deems it wise to omit material requiring calculus, I recommend using all or part of the following: Chapters 1, 2, 3 (omit Sections 3–3 and 3–4), 4, 5, 6, 7, 8 (omit Section 8–2), 9, 10 (omit Section 10–2 except for a brief discussion of Corollary 10–1), 11 (omit Section 11–2 save for a summary of the results), 12, and 15 (up to Definition 15–1).

George E. Andrews

CONTENTS

Part III ADDITIVITY

Part IV GEOMETRIC NUMBER THEORY

APPENDICES

Appendix B

Appendix C

Appendix D

PART I

MULTIPLICATIVITY-DIVISIBILITY

Part I is devoted to multiplicative problems; these are sometimes called divisibility problems, since division is the inverse of multiplication.

The knowledge of divisibility that we gain in the first two chapters leads us to our first goal, the fundamental theorem of arithmetic, which discloses the important role of primes in multiplicative number theory. Chapter 3 introduces combinatorial techniques for solving important divisibility problems and answering other number-theoretic questions. In order that we can study divisibility problems in greater depth, Chapters 4 and 5 develop the theory of congruences. Chapter 6 discusses some of the important functions related to multiplication and division, for example the number $d(n)$ of divisors of n and the sum $\sigma(n)$ of the divisors of n. Our results on congruences are extended in Chapter 7. The final chapter of Part I is concerned with the distribution of primes.

CHAPTER 1

BASIS REPRESENTATION

Our objective in this chapter is to prove the basis representation theorem (Theorem 1–3). First we need to understand the principle of mathematical induction, a tool indispensable in number theory.

1–1 PRINCIPLE OF MATHEMATICAL INDUCTION

Let us try to answer the following question: What is the sum of all integers from one through n, for any positive integer n? If $n=1$, the sum equals 1 because 1 is the only summand. The answer we seek is a formula that will enable us to determine this sum for each value of n without having to add the summands.

Table 1–1 lists the sum S_n of the first n consecutive positive integers for values of n from 1 through 10. Notice that in each case S_n equals one-half the product of n and the next integer; that is,

$$S_n = \frac{n(n+1)}{2} \tag{1-1-1}$$

for $n = 1,2,3,\ldots,10$. Although this formula gives the correct value of S_n for the first ten values of n, we cannot be sure that it holds for n greater than 10.

To construct Table 1–1, we do not need to compute S_n each time by adding the first n positive integers. Having obtained values of S_n

TABLE 1–1: SUM S_n OF THE FIRST n CONSECUTIVE POSITIVE INTEGERS.

n	S_n	n	S_n
1	1	6	21
2	3	7	28
3	6	8	36
4	10	9	45
5	15	10	55

for n less than or equal to some integer k, we can determine S_{k+1} simply by adding $(k+1)$ to S_k:

$$S_{k+1} = S_k + (k+1).$$

This last approach suggests a way of verifying equation (1–1–1). Suppose we know that formula (1–1–1) is true for $n \leq k$, where k is a positive integer. Then we know that

$$S_k = \frac{k(k+1)}{2},$$

and so

$$\begin{aligned} S_{k+1} &= S_k + (k+1) \\ &= \frac{k(k+1)}{2} + (k+1) \\ &= \left(\frac{k}{2} + 1\right)(k+1) \\ &= \frac{(k+2)(k+1)}{2}, \end{aligned}$$

that is,

$$S_{k+1} = \frac{(k+1)((k+1)+1)}{2}.$$

The last equation is the same as equation (1–1–1) except that n is replaced by $k+1$.

We have proved that if equation (1–1–1) holds for $n \leq k$, then it holds for $n = k+1$, and we have already verified that equation (1–1–1) holds for $n = 1,2, \ldots, 10$. Therefore, by the preceding argument, we conclude that equation (1–1–1) is also correct for $n = 11$. Since it holds for $n = 1,2, \ldots, 11$, the same process shows that it is correct for $n = 12$. Since it is true for $n = 1,2, \ldots, 12$, it is true for $n = 13$, and so on. We can describe the principle underlying the foregoing argument in various ways. The following formulation is the most appropriate for our purposes.

PRINCIPLE OF MATHEMATICAL INDUCTION: *A statement about integers is true for all integers greater than or equal to 1 if*

(i) *it is true for the integer 1, and*
(ii) *whenever it is true for all the integers $1,2, \ldots, k$, then it is true for the integer $k+1$.*

By "a statement about integers" we do not necessarily mean a formula. A sentence such as "$n(n^2-1)(3n+2)$ is divisible by 24" is also acceptable (see Exercise 17 of this section). The assumption that "the statement is true for $n = 1,2, \ldots, k$" will often be referred to as the *induction hypothesis*. Sometimes the role 1 plays in the Principle will be replaced by some other integer, say b; in such instances the principle of mathematical induction establishes the statement for all integers $n \geq b$.

Since we have shown that equation (1–1–1) fulfills conditions (i) and (ii), we conclude by the principle of mathematical induction that (1–1–1) is true for all integers $n \geq 1$. We state this result as our first theorem.

THEOREM 1–1: *The sum of the first n positive integers is* $\dfrac{n(n+1)}{2}$.

Our next theorem also illustrates the use of the principle of mathematical induction.

THEOREM 1–2: *If x is any real number other than* 1, *then* $\sum_{j=0}^{n-1} x^j = 1 + x + x^2 + \ldots + x^{n-1} = \dfrac{x^n - 1}{x - 1}$.

REMARK: $\sum_{j=0}^{n-1} A_j$ is shorthand for $A_0 + A_1 + A_2 + \ldots + A_{n-1}$.

PROOF: Again we proceed by mathematical induction. If $n = 1$, then $\sum_{j=0}^{1-1} x^j = x^0 = 1$ and $(x-1)/(x-1) = 1$. Thus the theorem is true for $n = 1$.

Assuming that $\sum_{j=0}^{k-1} x^j = (x^k - 1)/(x - 1)$, we find that

$$\sum_{j=0}^{(k+1)-1} x^j = \sum_{j=0}^{k-1} x^j + x^k = \frac{x^k - 1}{x - 1} + x^k$$

$$= \frac{x^k - 1 + x^{k+1} - x^k}{x - 1}$$

$$= \frac{x^{k+1} - 1}{x - 1}.$$

Hence condition (ii) is fulfilled, and we have established the theorem. ■

COROLLARY 1–1: *If m and n are positive integers and if* $m > 1$, *then* $n < m^n$.

PROOF: Note that

$$n = \underbrace{1 + 1 + 1 + \ldots + 1}_{n \text{ terms}} \le 1 + m + m^2 + \ldots + m^{n-1} = \frac{m^n - 1}{m - 1}$$

$$\le m^n - 1 < m^n. \quad \blacksquare$$

EXERCISES

1. Prove that

$$1^2 + 2^2 + 3^2 + \ldots + n^2 = \frac{n(n+1)(2n+1)}{6}.$$

2. Prove that

$$1^3 + 2^3 + 3^3 + \ldots + n^3 = (1 + 2 + 3 + \ldots + n)^2.$$

[Hint: Use Theorem 1–1.]

3. Prove that

$$x^n - y^n = (x - y)(x^{n-1} + x^{n-2}y + \ldots + xy^{n-2} + y^{n-1}).$$

4. Prove that

$$1 \cdot 2 + 2 \cdot 3 + 3 \cdot 4 + \ldots + n(n+1) = \frac{n(n+1)(n+2)}{3}.$$

5. Prove that

$$1 + 3 + 5 + \ldots + (2n - 1) = n^2.$$

6. Prove that

$$\frac{1}{2 \cdot 1} + \frac{1}{2 \cdot 3} + \frac{1}{3 \cdot 4} + \ldots + \frac{1}{n(n+1)} = \frac{n}{n+1}.$$

7. Suppose that $F_1 = 1$, $F_2 = 1$, $F_3 = 2$, $F_4 = 3$, $F_5 = 5$, and in general $F_n = F_{n-1} + F_{n-2}$ for $n \ge 3$. (F_n is called the nth Fibonacci number.) Prove that

$$F_1 + F_2 + F_3 + \ldots + F_n = F_{n+2} - 1.$$

In Exercises 8 through 16, F_n stands for the nth Fibonacci number. (See Exercise 7.)

8. Prove that

$$F_1 + F_3 + F_5 + \ldots + F_{2n-1} = F_{2n}.$$

9. Prove that

$$F_2 + F_4 + F_6 + \ldots + F_{2n} = F_{2n+1} - 1.$$

10. Prove that

$$F_{n+1}^2 - F_n F_{n+2} = (-1)^n.$$

11. Prove that

$$F_1F_2 + F_2F_3 + F_3F_4 + \ldots + F_{2n-1}F_{2n} = F_{2n}^2.$$

12. Prove that

$$F_1F_2 + F_2F_3 + F_3F_4 + \ldots + F_{2n}F_{2n+1} = F_{2n+1}^2 - 1.$$

13. The Lucas numbers L_n are defined by the equations $L_1 = 1$, and $L_n = F_{n+1} + F_{n-1}$ for each $n \geq 2$. Prove that

$$L_n = L_{n-1} + L_{n-2} \quad (n \geq 3).$$

14. What is wrong with the following argument?

"Assuming $L_n = F_n$ for $n = 1, 2, \ldots, k$, we see that

$$\begin{aligned} L_{k+1} &= L_k + L_{k-1} && \text{(by Exercise 13)} \\ &= F_k + F_{k-1} && \text{(by our assumption)} \\ &= F_{k+1} && \text{(by definition of } F_{k+1}). \end{aligned}$$

Hence, by the principle of mathematical induction, $F_n = L_n$ for each positive integer n."

15. Prove that $F_{2n} = F_n L_n$.

16. Prove that

$$L_1 + 2L_2 + 4L_3 + 8L_4 + \ldots + 2^{n-1}L_n = 2^n F_{n+1} - 1.$$

17. Prove that $n(n^2-1)(3n+2)$ is divisible by 24 for each positive integer n.

18. Prove that if n is an odd positive integer, then $x+y$ is a factor of x^n+y^n. (For example, if $n=3$, then $x^3+y^3=(x+y)(x^2-xy+y^2)$.)

1–2 THE BASIS REPRESENTATION THEOREM

Early in grade school, you learned to express the integers in terms of the decimal system of notation. The number ten is said to be the *base* for decimal notation, because the digits in any integer are coefficients of the progressive powers of 10.

Example 1–1: In the decimal system, two hundred nine is written 209, which stands for

$$2\cdot 10^2+0\cdot 10^1+9\cdot 10^0.$$

Similarly, for four thousand one hundred twenty-nine we write 4129, which stands for

$$4\cdot 10^3+1\cdot 10^2+2\cdot 10^1+9\cdot 10^0.$$

We can likewise express integers in binary, or base two, notation. In this case the digits 0 and 1 are used as the coefficients of the progressive powers of 2.

Example 1–2: In binary notation, we write twenty-three as 10111, which stands for

$$1\cdot 2^4+0\cdot 2^3+1\cdot 2^2+1\cdot 2^1+1\cdot 2^0,$$

and thirty-six has the form 100100, which stands for

$$1\cdot 2^5+0\cdot 2^4+0\cdot 2^3+1\cdot 2^2+0\cdot 2^1+0\cdot 2^0.$$

The basis representation theorem states that each integer greater than 1 can serve as a base for representing the positive integers.

THEOREM 1–3 *(Basis Representation Theorem): Let k be any integer larger than 1. Then, for each positive integer n, there exists a representation*

$$n=a_0k^s+a_1k^{s-1}+\ldots+a_s, \tag{1-2-1}$$

where $a_0 \neq 0$, and where each a_i is nonnegative and less than k. Furthermore, this representation of n is unique; it is called the representation of n to the base k.

REMARK: For each base k, we can also represent 0 by letting all the a_i be equal to 0.

PROOF: Let $b_k(n)$ denote the number of representations of n to the base k. We must show that $b_k(n)$ always equals 1.

It is possible that some of the coefficients a_i in a particular representation of n are equal to zero. Without affecting the representation, we may exclude terms that are zero. Thus suppose that

$$n = a_0k^s + a_1k^{s-1} + \ldots + a_{s-t}k^t,$$

where now neither a_0 nor a_{s-t} equals zero. Then

$$\begin{aligned} n-1 &= a_0k^s + a_1k^{s-1} + \ldots + a_{s-t}k^t - 1 \\ &= a_0k^s + a_1k^{s-1} + \ldots + (a_{s-t} - 1)k^t + k^t - 1 \\ &= a_0k^s + a_1k^{s-1} + \ldots + (a_{s-t} - 1)k^t + \sum_{j=0}^{t-1} (k-1)k^j, \end{aligned}$$

by Theorem 1–2 with $x = k$. Thus we see that for each representation of n to the base k, we can find a representation of $n-1$. If n has another representation to the base k, the same procedure will yield a new representation of $n-1$. Consequently

$$b_k(n) \leq b_k(n-1). \qquad (1\text{-}2\text{-}2)$$

It is important to note that inequality (1–2–2) holds even if n has *no* representation because $b_k(n) = 0 \leq b_k(n-1)$ in that case. Inequality (1–2–2) implies the following inequalities:

$$b_k(n+2) \leq b_k(n+1) \leq b_k(n),$$

$$b_k(n+3) \leq b_k(n+2) \leq b_k(n+1) \leq b_k(n),$$

and, in general, if $m \geq n+4$,

$$b_k(m) \leq b_k(m-1) \leq b_k(m-2) \leq \ldots \leq b_k(n+1) \leq b_k(n).$$

Since $k^n > n$ by Corollary 1–1, and since k^n clearly has at least one representation (namely, itself), we see that

$$1 \leq b_k(k^n) \leq b_k(n) \leq b_k(1) = 1.$$

The extreme entries in this set of inequalities are ones, so that all of the intermediate entries must be equal to 1. Thus $b_k(n) = 1$, and Theorem 1–3 is established. ■

Once a base $k (k > 1)$ has been chosen, we can represent any positive integer n uniquely as a sum of multiples of powers of k:

$$n = a_s k^s + a_{s-1} k^{s-1} + \ldots + a_1 k + a_0$$

where $a_0, a_1, \ldots, a_s$ stand for nonnegative integers less than k. This representation is usually denoted by the symbol "$a_s a_{s-1} \ldots a_1 a_0$" (not a product). For bases less than or equal to ten, the a_i are chosen from the symbols 0, 1, 2, . . ., 9 with their usual meanings; however, if k is greater than ten, one must invent additional symbols in order to have a total of k different symbols (one for each of the integers from zero to $k-1$).

Example 1–3: Let A stand for ten; B, for eleven; C, for twelve; D, for thirteen; E, for fourteen; and F, for fifteen. Using these symbols, we write three hundred to the base sixteen as $12C$, that is,

$$1 \cdot 16^2 + 2 \cdot 16^1 + 12 \cdot 16^0.$$

Similarly, two hundred is represented as $C8$; one hundred, as 64; and ten, as A.

This ability to represent integers to any base greater than one is much more than a mathematical curiosity. The bases 2, 8, and 16 are important in computer science. More useful to us, however, is the applicability of Theorem 1–3 in the proofs of many results about integers. We shall obtain some of these results in the next chapter.

EXERCISES

1. Write the numbers twenty-five, thirty-two, and fifty-six to the base five.

2. Write the numbers forty-seven, sixty-eight, and one hundred twenty-seven to the base 2.

3. What is the least number of weights required to weigh any integral number of pounds up to 63 pounds if one is allowed to put weights in only one pan of a balance?

4. Prove that each nonzero integer may be uniquely represented in the form

$$n = \sum_{j=0}^{s} c_j 3^j,$$

where $c_s \neq 0$, and each c_j is equal to -1, 0, or 1.

5. Using Exercise 4, determine the least number of weights required to weigh any integral number of pounds up to 80 pounds if one is allowed to put weights in both pans of a balance.

6. Prove that if

$$a_s k^s + a_{s-1} k^{s-1} + \ldots + a_0$$

is a representation of n to the base k, then $0 < n \leq k^{s+1} - 1$. [Hint: Use Theorem 1-2.]

7. Without using Theorem 1-3, prove directly that two different representations to the base k represent different integers. [Hint: Use Exercise 6.]

CHAPTER 2

THE FUNDAMENTAL THEOREM OF ARITHMETIC

In every branch of mathematics we meet theorems that seem so natural that, if we held no respect for logical rigor, we would be tempted to take them for granted. We must prove such theorems carefully, not only because they may be crucial in the logical structure of the theory, but also because every few years some proposition whose denial has long appeared to be utterly unacceptable to common sense turns out to be false.

You are now acquainted with one of these important theorems, the basis representation theorem (Theorem 1–3). This chapter will culminate in another basic proposition, the fundamental theorem of arithmetic (Theorem 2–5), from which we shall obtain significant information about the multiplicative structure of the integers. In passing, we note that a certain apparently obvious extension of the theorem to other number-theoretic structures resembling the integers is false (see Exercise 1 in Section 2–4).

We begin by developing Euclid's division lemma (Theorem 2–1), by means of which we shall study the divisibility properties of integers (Theorems 2–2 and 2–3). Knowledge of these properties will enable us to prove the fundamental theorem of arithmetic.

2–1 EUCLID'S DIVISION LEMMA

The division lemma furnishes the foundation for much of number theory; yet it is simply a rigorous restatement of the well-known fact that division of one integer by another yields an integral quotient and an integral nonnegative remainder smaller than the divisor. In order to avoid unnecessary complications, we limit ourselves to positive divisors. The proof we shall give for the lemma relies heavily on the basis representation theorem.

Theorem 2-1 *(Euclid's Division Lemma): For any integers k $(k > 0)$ and j, there exist unique integers q and r such that $0 \leq r < k$ and*

$$j = qk + r. \tag{2-1-1}$$

Proof: Note that we have simply rewritten a division problem in terms of multiplication and addition. In the notation used above, j is the dividend; k, the divisor; q, the quotient; and r, the remainder.

If $k = 1$, r must be zero, so that $q = j$.

If $k > 1$, suppose first that $j > 0$. (We shall consider the cases in which $j = 0$ and $j < 0$ later.) By the basis representation theorem (Theorem 1–3), j has a unique representation to the base k, say

$$\begin{aligned} j &= a_s k^s + a_{s-1} k^{s-1} + \ldots + a_1 k + a_0 \\ &= k(a_s k^{s-1} + a_{s-1} k^{s-2} + \ldots + a_1) + a_0 \\ &= kq + r, \end{aligned}$$

where $0 \leq r = a_0 < k$.

If a second pair q' and r' existed, we could find a representation for q' to the base k, say

$$q' = b_t k^t + \ldots + b_1 k + b_0,$$

so that

$$\begin{aligned} j &= kq' + r' \\ &= b_t k^{t+1} + \ldots + b_1 k^2 + b_0 k + r', \end{aligned}$$

but

$$j = a_s k^s + a_{s-1} k^{s-1} + \ldots + a_1 k + a_0.$$

By the uniqueness of the representation of j to the base k, we see that $t = s - 1$, $b_i = a_{i+1}$, $r' = a_0 = r$, and thus

$$\begin{aligned} q' &= b_t k^t + \ldots + b_1 k + b_0 \\ &= a_s k^{s-1} + \ldots + a_2 k + a_1 \\ &= q. \end{aligned}$$

Consequently, the theorem is true for positive values of j.

If $j = 0$, it is easy to verify that $q = r = 0$ is the only possible solution of (2–1–1) with $0 \leq r < k$.

If $j < 0$, then $-j > 0$, and there exist unique integers q'' and r'' such that

$$-j = kq'' + r''.$$

If $r'' = 0$, then $j = k(-q'')$; thus we may take $q = -q''$ and $r = 0$. If $r'' \neq 0$, then

$$\begin{aligned} j &= -kq'' - r'' \\ &= k(-q'' - 1) + (k - r''), \end{aligned}$$

and we may take $q = -q'' - 1$, and $r = k - r''$.

In either case, q and r satisfy equation (2–1–1). Uniqueness for negative j follows from uniqueness for $-j$, which is then positive. ■

EXERCISES

1. Without assuming Theorem 2–1, prove that for each pair of integers j and $k (k > 0)$, there exists some integer q for which $j - qk$ is positive.

2. The principle of mathematical induction is equivalent to the following statement, called the least-integer principle: *Every non-empty set of positive integers has a least element.*

   Using the least integer principle, define r to be the least integer for which $j - qk$ is positive (see Exercise 1). Prove that $0 < r \leq k$.

3. Use Exercise 2 to give a new proof of Theorem 2–1.

4. Any nonempty set of integers J that fulfills the following two conditions is called an *integral ideal:*

   (i) if n and m are in J, then $n + m$ and $n - m$ are in J; and
   (ii) if n is in J and r is an integer, then rn is in J.

   Let $\mathcal{J}_m$ be the set of all integers that are integral multiples of a particular integer m. Prove that $\mathcal{J}_m$ is an integral ideal.

5. Prove that every integral ideal J is identical with $\mathcal{J}_m$ for some m. [Hint: Prove that if $J \neq \{0\} = \mathcal{J}_0$, then there exist positive integers in J. By the least-integer principle

(Exercise 2), there is a least positive integer in J, say m. Then prove that $J = \mathscr{I}_m$.]

6. Prove that if a and b are odd integers, then $a^2 - b^2$ is divisible by 8.

7. Prove that if a is an odd integer, then $\{a^2 + (a+2)^2 + (a+4)^2 + 1\}$ is divisible by 12.

2-2 DIVISIBILITY

If a and $b\,(b \neq 0)$ are integers, we say *b divides a*, or *b is a divisor of a*, if a/b is an integer. We shall write $b \mid a$ to indicate that b divides a; and, $b \nmid a$, to indicate that b does not divide a.

Example 2-1: $2 \mid 4$, but $3 \nmid 4$.

Example 2-2: If a is an integer, then $1 \mid a$ and $-1 \mid a$; furthermore, if $a \neq 0$, then $a \mid a$ and $-a \mid a$.

Example 2-3: For each nonzero integer a, $a \mid 0$.

Example 2-4: Let a, b, c, and d be integers. Suppose that an integer e divides both a and c. Then there exist integers x and y such that $a = ex$ and $c = ey$. Therefore,

$$\begin{aligned} ab + cd &= exb + eyd \\ &= e(xb + yd), \end{aligned}$$

which implies that $e \mid (ab + cd)$. Consequently, if $e \mid a$ and $e \mid c$, then $e \mid (ab + cd)$ also.

If a and b are integers, then any integer that divides both a and b is called a *common divisor* of a and b.

DEFINITION 2-1: *If a and b are integers, not both zero, then an integer d is called the* **greatest common divisor** *of a and b if*

(i) $d > 0$,
(ii) *d is a common divisor of a and b, and*
(iii) *each integer f that is a common divisor of both a and b is also a divisor of d.*

We shall prove shortly that each pair of integers a and b, not both zero, has a unique greatest common divisor; this integer is denoted by g.c.d.(a,b). Many authors write (a,b) for g.c.d.(a,b). We do not, because we shall often use (a,b) to represent a point in the Euclidean plane.

Example 2–5: The positive divisors of 12 are 1, 2, 3, 4, 6, and 12. The positive divisors of -8 are 1, 2, 4, and 8. Thus the positive common divisors of 12 and -8 are 1, 2, and 4; hence, g.c.d.$(12,-8)=4$.

Example 2–6: If $a \neq 0$ and $a \mid b$, then g.c.d.$(a,b) = |a|$.

Our proof of the existence and uniqueness of the greatest common divisor depends completely on the Euclidean algorithm, a device involving nothing more than repeated application of the division lemma. Before proceeding with the proof, we illustrate the Euclidean algorithm with the following example.

Example 2–7: What is g.c.d.(341,527)? Dividing 341 into 527, we find that the q and r, as in Theorem 2–1, are 1 and 186, respectively, because

$$527 = 341 \cdot 1 + 186 \tag{2-2-2}$$

Clearly, any number that divides both 527 and 341 also divides 186; for, if $dc = 527$ and $de = 341$, then $d(c-e) = 186$.

In the same manner,

$$341 = 186 \cdot 1 + 155, \tag{2-2-3}$$

$$186 = 155 \cdot 1 + 31, \text{ and} \tag{2-2-4}$$

$$155 = 31 \cdot 5. \tag{2-2-5}$$

By equation (2–2–5), 31 divides 155. Therefore 31 divides 186, by (2–2–4); $31 \mid 341$, by (2–2–3); and $31 \mid 527$, by (2–2–2). Thus 31 satisfies (i) and (ii) in Definition 2–1. Finally, if $f \mid 341$ and $f \mid 527$, then $f \mid 186$, by (2–2–2); $f \mid 155$, by (2–2–3); and $f \mid 31$, by (2–2–4). Since all three conditions in Definition 2–1 are satisfied, we see that $31 =$ g.c.d.(341,527).

The proof of Theorem 2 involves nothing more than the procedure of Example 2–7 in a general setting.

Theorem 2–2: *If a and b are integers, not both zero, then* g.c.d.(a,b) *exists and is unique.*

PROOF: Clearly g.c.d.(a,b) is not affected by the signs of a and b. We have asserted that not both a and b are zero; however, if either is zero, say $b=0$, then g.c.d.$(a,0)$ is clearly equal to $|a|$. In the following proof, we may therefore assume that $a \geq b > 0$.

By Theorem 2–1, there exist q_1 and r_1 $(0 \leq r_1 < b)$ such that

$$a = bq_1 + r_1.$$

If $r_1 > 0$, there exist q_2 and r_2 such that

$$b = r_1q_2 + r_2,$$

where $0 \leq r_2 < r_1$. If $r_2 > 0$, there exist q_3 and r_3 such that

$$r_1 = r_2q_3 + r_3,$$

where $0 \leq r_3 < r_2$. This process may be continued as long as the newly arising r_i does not equal zero.

Since

$$b > r_1 > r_2 > r_3 > \ldots > 0,$$

we see, by mathematical induction, that $0 \leq r_i \leq b - i$. Therefore, in at most b steps, we shall obtain an r_n that is zero.

Thus the last application of Theorem 2–1 in our procedure leads to the result

$$r_{n-2} = r_{n-1}q_n + 0;$$

that is, $r_n = 0$. The computation of g.c.d.(341,527) in Example 2–7 suggests that r_{n-1} is equal to g.c.d.(a,b).

We have constructed the r_i so that $r_{n-1} > 0$. By working backward from the final equation, we may establish successively that r_{n-1} divides $r_{n-2}, r_{n-3}, \ldots, r_2, r_1, b$, and a. Finally, if f divides both a and b, we may proceed successively from the initial equation to deduce that f divides $r_1, r_2, \ldots, r_{n-2}$, and r_{n-1}. (Mathematical induction is tacitly used in both of these procedures.) Thus r_{n-1} satisfies the requirements of Definition 2–1; therefore, $r_{n-1} =$ g.c.d.(a,b).

Each pair of integers has only one greatest common divisor; for, if both d_1 and d_2 are greatest common divisors of some pair a and b, it follows from (iii) of Definition 2–1 that $gd_1 = d_2$ and $hd_2 = d_1$, where h and g are positive integers; hence, $d_2 = ghd_2$; thus $1 = gh$, and so $g = h = 1$. We conclude that $d_1 = d_2$. ■

An *integral linear combination* of the integers a and b is an expression of the form $ax + by$, where x and y are integers. We shall prove two corollaries of Theorem 2–2 that characterize those integers expressible as integral linear combinations of a particular pair of integers. First we consider an example.

Example 2–8: Using the results in Example 2–7, we shall express $31 = \text{g.c.d.}(341,527)$ as an integral linear combination of 341 and 527. We start with the next-to-the-last equation and successively substitute the other equations into it until we reach the initial equation. Equation (2–2–3) may be rewritten as

$$31 = 186 - 155 \cdot 1.$$

Using equation (2–2–3) to express 155, we find that

$$31 = 186 - (341 - 186 \cdot 1),$$

that is,

$$31 = 2 \cdot 186 - 341.$$

Using equation (2–2–2) to express 186, we see that

$$31 = 2 \cdot (527 - 341 \cdot 1) - 341,$$

that is,

$$31 = 2 \cdot 527 - 3 \cdot 341.$$

Thus we have expressed 31 as a linear combination of 341 and 527.

Note that, in addition,

$$31 = 14 \cdot 341 - 9 \cdot 527,$$

and

$$31 = -20 \cdot 341 + 13 \cdot 527.$$

In general, there may be many pairs x and y such that

$$\text{g.c.d.}(a,b) = ax + by.$$

Corollary 2–1: *If $d = \text{g.c.d.}(a,b)$, then there exist integers x and y such that*

$$ax + by = d. \tag{2-2-6}$$

PROOF: By taking the n equations used in the proof of Theorem 2–2 and using the principle of mathematical induction, we shall first establish that there exist integers x_i and y_i such that

$$ax_i + by_i = r_i \tag{2-2-7}$$

for $i = 1, 2, \ldots, n-1$.

When $i = 1$, let $x_1 = 1$, and $y_1 = -q_1$. Now assume that integer solutions of (2–2–7) have been found for all i less than or equal to $k(k < n-1)$. We know that

$$r_{k-1} = r_k q_{k+1} + r_{k+1}; \tag{2-2-8}$$

thus, by the induction hypothesis,

$$(ax_{k-1} + by_{k-1}) - (ax_k + by_k)q_{k+1} = r_{k+1}. \tag{2-2-9}$$

We can rewrite equation (2–2–9) in the form

$$(x_{k-1} - x_k q_{k+1})a + (y_{k-1} - y_k q_{k+1})b = r_{k+1}. \tag{2-2-10}$$

Hence, $x_{k+1} = x_{k-1} - x_k q_{k+1}$ and $y_{k+1} = y_{k-1} - y_k q_{k+1}$ are solutions of equation (2–2–7) when $i = k+1$.

Thus formula (2–2–7) is established for $i = 1, 2, \ldots, n-1$, by the principle of mathematical induction. In particular, if $i = n-1$ in equation (2–2–7), we get the relation

$$ax_{n-1} + by_{n-1} = r_{n-1} = \text{g.c.d.}(a,b). \qquad \blacksquare$$

COROLLARY 2–2: *In order that there exist integers x and y satisfying the equation*

$$ax + by = c, \tag{2-2-11}$$

it is necessary and sufficient that $d \mid c$, where $d = \text{g.c.d.}(a,b)$.

PROOF: Let $a = ed$, $b = fd$. Then, if (2–2–11) holds, we get the relation

$$c = edx + fdy = d(ex + fy).$$

Thus $d \mid c$.

On the other hand, if $d \mid c$, let $kd = c$. Then, by Corollary 2–1, there exist x' and y' such that

$$ax' + by' = d.$$

Hence

$$a(x'k) + b(y'k) = dk = c.$$

Thus $x = x'k$ and $y = y'k$ provide a solution of (2–2–11). ■

Our next theorem follows from Corollary 2–2; it will be the principal tool we shall use in our proof of the fundamental theorem of arithmetic. First we need some further definitions.

DEFINITION 2–2: *A positive integer p other than 1 is said to be a* **prime** *if its only positive divisors are* 1 and p.

The first few primes are 2, 3, 5, 7, 11, ... (Although the 1968 World Almanac lists 1 as a prime, it is convenient not to do so in number theory. As you will see, the statement of the fundamental theorem of arithmetic would be needlessly complicated if 1 were considered prime. Perhaps this fact has been impressed on the editors of the Almanac, for the 1969 and later editions do not list 1 as a prime.) The primes have many interesting properties, some of which we shall explore in later sections.

DEFINITION 2–3: *We say that a and b are* **relatively prime** *if* g.c.d.$(a,b) = 1$.

Example 2–9: The positive divisors of 7 are 1 and 7. The positive divisors of 27 are 1, 3, 9, and 27. Since 1 is the only positive common divisor of 7 and 27, these two integers are relatively prime.

Example 2–10: If $d =$ g.c.d.(a,b), then a/d and b/d are relatively prime. To show this, let x and y be integers such that $ax + by = d$. Then $(a/d)x + (b/d)y = 1$, and so g.c.d.$(a/d, b/d) = 1$.

Example 2–11: If p is a prime and a is an integer such that $p \nmid a$, then p and a are relatively prime. In particular, any two different primes are relatively prime.

THEOREM 2–3: *If a, b, and c are integers, where a and c are relatively prime, and if $c \mid ab$, then c divides b.*

PROOF: Since g.c.d.$(a,c) = 1$, Corollary 2–2 implies that there exist integers x and y such that

$$cx + ay = 1.$$

Therefore,

$$cbx + aby = b. \qquad (2\text{-}2\text{-}12)$$

Since $c \mid ab$, there exists a k such that $ab = kc$.

Substituting kc for ab in equation (2–2–12), we find that

$$cbx + kcy = b. \qquad (2\text{-}2\text{-}13)$$

Thus

$$c(bx + ky) = b. \qquad (2\text{-}2\text{-}14)$$

Hence $c \mid b$. ■

COROLLARY 2–3: *If a and b are integers, p is a prime, $p \mid ab$, and $p \nmid a$, then $p \mid b$.*

PROOF: If $p \nmid a$, then g.c.d.$(a,p) = 1$, because the only positive divisors of p are 1 and p. Hence, by Theorem 2–3 (with $c = p$), we see that $p \mid b$. ■

COROLLARY 2–4: *If $p \mid a_1 a_2 \ldots a_n$, then there exists some i such that $p \mid a_i$.*

PROOF: We proceed by mathematical induction. The assertion is clear for $n = 1$. For $n = 2$, it is a restatement of Corollary 2–3. We assume that the assertion is true for n less than or equal to k. Then for $n = k + 1$ we consider the relation

$$p \mid (a_1 a_2 \ldots a_k) a_{k+1}.$$

By Corollary 2–3, either $p \mid a_{k+1}$ (so that $i = k + 1$) or $p \mid a_1 a_2 \ldots a_k$, in which case $p \mid a_i$ for some $i (1 \le i \le k)$, by the induction hypothesis. ■

EXERCISES

1. Using the technique described in Example 2–7, find the greatest common divisor of the following pairs of integers.

   (a) 527, 765 (d) 108, 243

   (b) 361, 1178 (e) 132, 473

   (c) 12321, 8658 (f) 156, 1740.

2. Using the technique described in Example 2–8, find the greatest common divisor d of 299 and 481. Then find integers x and y such that

$$299x + 481y = d.$$

3. In Exercise 2, replace 299 and 481 by 129 and 301 and proceed as indicated.

4. The *least common multiple* of two positive integers a and b (denoted by l.c.m.(a,b)) is defined to be the smallest positive integer that is divisible by both a and b. Prove that

$$\text{l.c.m.}(a,b) = \frac{ab}{\text{g.c.d.}(a,b)}.$$

5. Compute the following:

   (a) l.c.m.(25,30) (d) l.c.m.(28,29)

   (b) l.c.m.(42,49) (e) l.c.m.$(n,n+1)$

   (c) l.c.m.(27,81) (f) l.c.m.$(2n-1,2n+1)$.

6. Prove that l.c.m.$(ab,ad) = a[\text{l.c.m.}(b,d)]$.

7. Prove that if $D = d/\text{g.c.d.}(b,d)$ and $B = b/\text{g.c.d.}(b,d)$, then

$$\frac{a}{b} + \frac{c}{d} = \frac{aD + cB}{\text{l.c.m.}(b,d)}.$$

   Discuss the relationship between this equation and the addition of fractions by means of a "common denominator".

8. Prove that g.c.d.$(a+b, a-b) \geq$ g.c.d.(a,b).

9. Prove that, if a and b are nonzero integers, then g.c.d.$(a,b) \mid$ l.c.m.(a, b).

10. Let $\mathscr{I}_m$ be the set of all integral multiples of the integer m. Prove that

$$\mathscr{I}_m \cap \mathscr{I}_n = \mathscr{I}_{\text{l.c.m.}(m,n)}.$$

   [If S and T are sets, then $S \cap T$ denotes the set of elements common to both S and T].

11. Prove that $\mathscr{I}_{\text{g.c.d.}(m,n)}$ contains all the elements of $\mathscr{I}_m$ and all the elements of $\mathscr{I}_n$. Prove that if $\mathscr{I}_r$ contains all the elements of $\mathscr{I}_m$ and $\mathscr{I}_n$, then $\mathscr{I}_r$ contains all the elements of $\mathscr{I}_{\text{g.c.d.}(m,n)}$.

12. We can define a *generalized Fibonacci sequence* $\mathscr{F}_1$, $\mathscr{F}_2$, $\mathscr{F}_3$, $\mathscr{F}_4$, ... by first selecting four integers a, b, c, and e, and then letting $\mathscr{F}_1 = a$, $\mathscr{F}_2 = b$, and $\mathscr{F}_n = c\mathscr{F}_{n-1} + e\mathscr{F}_{n-2}$ if $n > 2$.

    (a) Prove that, if $d = \text{g.c.d.}(a,b)$, then $d \mid \mathscr{F}_n$ for all $n \geq 1$.

    (b) Prove that, if $f = \text{g.c.d.}(\mathscr{F}_m, \mathscr{F}_{m-1})$ and $\text{g.c.d.}(f, e) = 1$, then $f \mid d$.

2-3 THE LINEAR DIOPHANTINE EQUATION

We have now amassed enough results to prove the fundamental theorem of arithmetic. Before beginning this task, however, we shall consider a result related to Corollary 2-2.

Let a, b, and c be integers ($a \neq 0 \neq b$). The expression

$$ax + by = c \tag{2-3-1}$$

is called a *linear Diophantine equation*. A *solution* of this equation is a pair (x,y) of integers that satisfies the equation.

From analytic geometry we know that each point in a plane can be associated with an ordered pair of real numbers called coordinates. A point whose coordinates are integers is called a *lattice point*. In the plane, the locus of points whose coordinates x and y satisfy equation (2-3-1) is a straight line. Thus the solutions of this linear Diophantine equation correspond to the lattice points lying on the straight line. Depending on the values of a, b, and c, there may be none or many lattice points on the graph of $ax + by = c$.

From Corollary 2-2 we know that the linear Diophantine equation $ax + by = c$ has a solution if and only if $d \mid c$, where $d = \text{g.c.d.}(a,b)$. Suppose that d *does* divide c. Using the procedure illustrated in Example 2-8, we can find w_0 and z_0 such that

$$aw_0 + bz_0 = d.$$

Next, we find an integer k such that $c = dk$; and we let $x_0 = w_0k$ and $y_0 = z_0k$. Clearly, (x_0, y_0) is a solution of equation (2-3-1). Suppose (x', y') is also a solution of equation (2-3-1). Then

$$ax' + by' = c = ax_0 + by_0,$$

and so

$$\frac{a}{d}x' + \frac{b}{d}y' = \frac{a}{d}x_0 + \frac{b}{d}y_0.$$

Therefore,

$$\frac{a}{d}(x' - x_0) = \frac{b}{d}(y_0 - y'). \qquad \text{(2-3-2)}$$

By Example 2–10, g.c.d.$(a/d, b/d) = 1$; thus, by Theorem 2–3,

$$\frac{b}{d} \,\Big|\, (x' - x_0).$$

Hence, there exists an integer t such that $x' - x_0 = tb/d$; that is, $x' = x_0 + tb/d$. Substituting tb/d for $x' - x_0$ in equation (2–3–2), we find that

$$\frac{a}{d}t\frac{b}{d} = \frac{b}{d}(y_0 - y'),$$

and so

$$y_0 - y' = t\frac{a}{d},$$

that is,

$$y' = y_0 - t\frac{a}{d}.$$

We conclude that, for each solution (x', y') of equation (2–3–1), there exists an integer t such that

$$x' = x_0 + t\frac{b}{d} \qquad \text{and} \qquad y' = y_0 - t\frac{a}{d}.$$

In fact, $(x_0 + tb/d,\ y_0 - ta/d)$ is a solution of equation (2–3–1) for each t, because

$$a\left(x_0 + t\frac{b}{d}\right) + b\left(y_0 - t\frac{a}{d}\right) = ax_0 + by_0 + t\frac{ab}{d} - t\frac{ab}{d} = c.$$

We now summarize the preceding results.

Theorem 2–4: *The linear Diophantine equation*

$$ax + by = c$$

has a solution if and only if $d \mid c$, where $d =$ g.c.d.(a,b). Furthermore, *if (x_0,y_0) is a solution of this equation, then the set of solutions of the equation consists of all integer pairs (x,y), where*

$$x = x_0 + t\frac{b}{d} \quad \textit{and} \quad y = y_0 - t\frac{a}{d} \qquad (t = \ldots, -2, -1, 0, 1, 2, \ldots).$$

Example 2–12: The linear Diophantine equation $15x + 27y = 1$ has no solutions, since g.c.d.$(15,27) = 3$ and $3 \nmid 1$.

Example 2–13: The linear Diophantine equation $5x + 6y = 1$ has a solution, since g.c.d.$(5,6) = 1$. By inspection, we see that $(-1,1)$ is such a solution. Hence, all solutions are given by (x,y), where $x = -1 + 6t$, $y = 1 - 5t$ $(t = \ldots -2, -1, 0, 1, 2, \ldots)$.

EXERCISES

1. Find the general solution (if solutions exist) of each of the following linear Diophantine equations:

   (a) $2x + 3y = 4$
   (b) $17x + 19y = 23$
   (c) $15x + 51y = 41$
   (d) $23x + 29y = 25$
   (e) $10x - 8y = 42$
   (f) $121x - 88y = 572$.

2. A man pays \$1.43 for some apples and pears. If pears cost 17¢ each, and apples, 15¢ each, how many of each did he buy?

3. Draw the graphs of the straight lines defined by the equations in parts (a), (b), and (c) of Exercise 1.

4. Prove that the area of the triangle whose vertices are $(0,0)$, (b,a), and (x,y) is $|by - ax|/2$.

5. Prove that if (x_0,y_0) is a solution of the linear Diophantine equation $ax - by = 1$, then the area of the triangle whose vertices are $(0,0)$, (b,a), and (x_0,y_0) is $1/2$.

6. Is there a nondegenerate triangle with area smaller than $1/2$ and with vertices (p_1,q_1), (p_2,q_2), and (p_3,q_3), where the p_i and q_i are integers? Prove your answer.

7. What is the perpendicular distance to the origin $(0,0)$ from the line defined by the equation

$$ax - by = 1?$$

8. What is the shortest possible distance between two lattice points on the line defined by the linear Diophantine equation

$$ax - by = c?$$

(Recall that, by the definition of a linear Diophantine equation, the constants a, b, and c must be integers.)

2-4 THE FUNDAMENTAL THEOREM OF ARITHMETIC

Table 2-1 exhibits the ways the first twelve positive integers may be factored into primes.

The evidence of Table 2-1 suggests that there is exactly one prime factorization of each integer greater than 1, if the order of the prime factors is disregarded.

While not as intuitively apparent as the basis representation theorem (Theorem 1-3), the foregoing conjecture not only is true, but is so important to the study of integers that it is called the fundamental theorem of arithmetic.

THEOREM 2-5 *(Fundamental Theorem of Arithmetic): For each integer $n > 1$, there exist primes $p_1 \leq p_2 \leq p_3 \leq \ldots \leq p_r$ such that*

$$n = p_1 p_2 \ldots p_r;$$

this factorization is unique.

PROOF: Our first goal is to prove that each integer has at least one prime factorization. Note that (see Table 2-1) such a factorization

TABLE 2-1: FACTORIZATION OF INTEGERS INTO PRIMES.

n	factorizations
1	—
2	2
3	3
4	$2 \cdot 2$
5	5
6	$2 \cdot 3 = 3 \cdot 2$
7	7
8	$2 \cdot 2 \cdot 2$
9	$3 \cdot 3$
10	$2 \cdot 5 = 5 \cdot 2$
11	11
12	$2 \cdot 2 \cdot 3 = 2 \cdot 3 \cdot 2 = 3 \cdot 2 \cdot 2$

exists for all $n(2 \le n \le 12)$. Let us now assume that each integer $m(1 < m \le k)$ can be factored into primes.

Now, either $k+1$ is prime or it is not. If it is prime, then its prime factorization consists just of the prime itself. If $k+1$ is not prime, then

$$k+1 = ab,$$

where $1 < a < k+1$ and $1 < b < k+1$. Since $1 < a \le k$ and $1 < b \le k$, both a and b have prime factorizations, say

$$a = p_1 p_2 \ldots p_s \qquad \text{and} \qquad b = p_1' p_2' \ldots p_t'.$$

Therefore,

$$k+1 = p_1 p_2 \ldots p_s p_1' p_2' \ldots p_t'.$$

Hence $k+1$ has a prime factorization. Thus we have established by mathematical induction that every integer greater than 1 has a prime factorization.

To complete the theorem, we must establish uniqueness of factorization. Again we proceed by mathematical induction. Our table also tells us that the factorization of each n ($n \le 12$) is unique. Assume that each integer $m(1 < m \le k)$ has a unique prime factorization. Suppose that $k+1$ has the two prime factorizations

$$k+1 = p_1 p_2 \ldots p_u = p_1' p_2' \ldots p_v',$$

where $p_1 \le p_2 \le \ldots \le p_u$ and $p_1' \le p_2' \le \ldots \le p_v'$. Since p_1' divides $k+1$, we see that p_1' divides $p_1 p_2 \ldots p_u$; thus p_1' divides p_i for some i, by Corollary 2–4. Since both p_1' and p_i are primes, we conclude that $p_1' = p_i$.

Clearly, we may reverse the preceding argument to show that $p_1 = p_j'$ for some j. Hence

$$p_1 = p_j' \ge p_1',$$

and

$$p_1' = p_i \ge p_1.$$

Therefore, $p_1 \ge p_1' \ge p_1$; and so $p_1 = p_1'$. Thus $(k+1)/p_1$ is an integer not exceeding k, and

$$p_2 \ldots p_u = \frac{k+1}{p_1} = p_2' \ldots p_v'.$$

Hence, by the induction hypothesis, $u = v$, $p_2' = p_2, \ldots,$ and $p_u' = p_u$. Thus the fundamental theorem of arithmetic is established. ■

EXERCISES

1. Let E be the set of all positive even integers. Define m to be an "even prime" if m is even but is not factorable into two even numbers. Prove that some elements of E are not uniquely representable as products of even primes.

2. Prove that every positive integer is uniquely representable as the product of a nonnegative power of 2 (perhaps 2^0) and an odd number.

3. Suppose that $a = p_1 p_2 \ldots p_s$ is the unique factorization of a into primes ($p_1 \leq p_2 \leq \ldots \leq p_s$). Prove that a has a unique representation

$$q_1^{e_1} q_2^{e_2} \ldots q_r^{e_r},$$

where the q_i are primes, $q_1 < q_2 < \ldots < q_r$, and the e_i are positive integers.

4. Prove that, if $a = q_1^{e_1} q_2^{e_2} \ldots q_r^{e_r}$ and $b = s_1^{f_1} s_2^{f_2} \ldots s_u^{f_u}$ are the factorizations of a and b into primes (see Exercise 3), then there exist primes $t_1 < t_2 < \ldots < t_v$ and nonnegative integers g_i and h_i such that

$$a = t_1^{g_1} t_2^{g_2} \ldots t_v^{g_v}, \text{ and } b = t_1^{h_1} t_2^{h_2} \ldots t_v^{h_v}.$$

5. Using the notation of Exercise 4, prove that

$$\text{g.c.d.}(a,b) = t_1^{c_1} t_2^{c_2} \ldots t_v^{c_v},$$

where each c_i is the smaller of the corresponding g_i and h_i.

6. Use Exercise 5 to find

(a) g.c.d.(121,66)
(b) g.c.d.(169,273)
(c) g.c.d.(51,187)
(d) g.c.d.(2187,999)
(e) g.c.d.(64,81)
(f) g.c.d.(p^2q,pqr), where p, q, and r are primes.

7. Using the notation of Exercise 4, prove that

$$\text{l.c.m.}(a,b) = t_1^{j_1} t_2^{j_2} \ldots t_v^{j_v},$$

where each j_i is the largest of the corresponding g_i and h_i.

8. Do Exercise 4 of Section 2–2, using Exercises 5 and 7 of this section.

9. Use Exercise 7 to find

   (a) l.c.m.(125,150)

   (b) l.c.m.(132,154)

   (c) l.c.m.(39,143)

   (d) l.c.m.(253,506)

   (e) l.c.m.(111,1221)

   (f) l.c.m.(p^2q,pqr), where p, q, and r are primes.

10. For each finite set of integers $\{a,b,c,\ldots,r\}$, we can define

$$\text{g.c.d.}(a,b,c,\ldots,r)$$

to be the largest integer that divides each of a, b, c, . . ., and r. We can also define

$$\text{l.c.m.}(a,b,c,\ldots,r)$$

as the smallest integer that is divisible by each of a, b, c, . . ., and r. Find formulae for g.c.d.$(a,b,c,\ldots,r)$ and l.c.m.$(a,b,c,\ldots,r)$ by generalizing the assertions in Exercises 4, 5, and 7.

11. Find g.c.d.(39,102,75) and l.c.m.(39,102,75).

12. Prove that, if $d_1 = \text{g.c.d.}(a,b)$, $d_2 = \text{g.c.d.}(b,c)$, $d_3 = \text{g.c.d.}(c,a)$, $D = \text{g.c.d.}(a,b,c)$, and $L = \text{l.c.m.}(a,b,c)$, then

$$L = \frac{abcD}{d_1d_2d_3}.$$

CHAPTER 3

COMBINATORIAL AND COMPUTATIONAL NUMBER THEORY

Much of number theory is concerned with the properties of primes. In Chapter 2 we saw that these numbers are the multiplicative building blocks of the integers. In Sections 3–2 and 3–3, we shall use combinatorial techniques to obtain two surprising results about primes. The combinatorial ideas underlying this approach will also be used in proving many of the theorems in later chapters. In the fourth section, we shall introduce one of number theory's most useful tools, the generating function. To conclude the chapter, we shall discuss the role of computers in number theory.

3–1 PERMUTATIONS AND COMBINATIONS

Although permutations and combinations are usually associated with probability theory, they are also relevant to number theory. For instance, let us consider a problem that faces a number theorist each time he visits a Chinese restaurant.

Example 3–1: The Dinners for Two on a particular Chinese menu are presented as follows:

DINNERS FOR TWO	
You may select one dish from each category.	
Category A	Category B
Fung Wong Guy	Chicken Chow Mein
Wor Hip Har	Ho Yu Gai Poo
Moo Goo Guy Pen	

How many different Dinners for Two are available? Without any difficulty, we can list all the available dinners:

Fung Wong Guy and Chicken Chow Mein,
Fung Wong Guy and Ho Yu Gai Poo,
Wor Hip Har and Chicken Chow Mein,
Wor Hip Har and Ho Yu Gai Poo,
Moo Goo Guy Pen and Chicken Chow Mein, and
Moo Goo Guy Pen and Ho Yu Gai Poo.

Of course, we may easily count the dinners without listing them. We have 3 choices in Category A, and, after we make a decision there, we have 2 choices in Category B. Thus, without looking at the list, we note that there are

$$2+2+2=3\cdot 2=6$$

different dinners.

The simple counting procedure employed in Example 3–1 is a particular instance of the following fundamental rule.

General Combinatorial Principle: *If an element α can be chosen from a prescribed set S in m different ways, and if thereafter, a second element β can be chosen from a prescribed set T in n different ways, then the ordered pair (α,β) can be chosen in mn different ways.**

You may be wondering what all this can *really* have to do with number theory. The following examples lead us to expect that the product of any n consecutive positive integers is divisible by the product of the first n positive integers; though this assertion appears to have no direct relationship to combinatorial ideas, we shall see that the proof of it involves all the combinatorial concepts to be introduced in this section.

Example 3–2: For $n=4$, the product of the first four integers is $1\cdot 2\cdot 3\cdot 4=24$, and we observe that $5\cdot 6\cdot 7\cdot 8=1680=70\cdot 24$; also $10\cdot 11\cdot 12\cdot 13=17160=715\cdot 24$.

Example 3–3: For $n=5$, the product of the first 5 integers is $1\cdot 2\cdot 3\cdot 4\cdot 5=120$, and we observe that $4\cdot 5\cdot 6\cdot 7\cdot 8=6720=56\cdot 120$; also $11\cdot 12\cdot 13\cdot 14\cdot 15=360360=3003\cdot 120$.

*This principle is actually a theorem in the foundations of mathematics. See Theorem 10.4.12 in *The Anatomy of Mathematics* by Kershner and Wilcox.

DEFINITION 3–1: *An* **r-permutation** *of a set S of n objects is an ordered selection of r elements from S.*

Example 3–4: If $S = \{4,5,6\}$, then the 2-permutations of this set are (4,5), (5,4), (4,6), (6,4), (5,6), and (6,5); the 3-permutations of this set are (4,5,6), (4,6,5), (5,4,6), (5,6,4), (6,5,4), and (6,4,5).

THEOREM 3–1: *If ${}_nP_r$ denotes the number of r-permutations of a set of n objects, then*

$${}_nP_r = n(n-1)\ldots(n-r+1).$$

PROOF: We may make our first selection of an object in n different ways. Once the first selection has been made, we may make the second selection in $n-1$ ways from the remaining elements of the set. By mathematical induction, we see that we may make the ith selection in exactly $n-(i-1)$ ways. Thus the general combinatorial principle tells us that the r selections we are to make can be executed in

$$n(n-1)(n-2)\ldots(n-r+1)$$

different ways. Thus we obtain the desired formula

$${}_nP_r = n(n-1)\ldots(n-r+1). \qquad \blacksquare$$

Of special interest to us is the symbol ${}_rP_r$, which we shall write as $r!$ (read r factorial); we observe that $1! = 1, 2! = 2, 3! = 6, 4! = 24$, and so on.

DEFINITION 3–2: *An* **r-combination** *of a set S of n objects is a subset of S having r elements.*

Example 3–5: If $S = \{4,5,6\}$, then the 2-combinations of S *are* $\{4,5\}, \{4,6\}$, and $\{5,6\}$.

Note that r-permutations include all possible orderings and are therefore more numerous than r-combinations.

THEOREM 3–2: *If $\binom{n}{r}$ denotes the number of r-combinations taken from a set S of n elements, then*

$$\binom{n}{r} = \frac{n(n-1)\ldots(n-r+1)}{r!}.$$

PROOF: To each of the $\binom{n}{r}$ different r-combinations of S, we may give ${}_rP_r$ different orderings. Consequently,

$$ {}_rP_r \cdot \binom{n}{r} = {}_nP_r. \qquad (3\text{-}1\text{-}1)$$

Thus,

$$\binom{n}{r} = \frac{{}_nP_r}{{}_rP_r} = \frac{n(n-1)\ldots(n-r+1)}{r!}. \qquad \blacksquare$$

We are now able to prove the conjecture made near the beginning of this section.

THEOREM 3–3: *The product of any n consecutive positive integers is divisible by the product of the first n positive integers.*

PROOF: The product of n consecutive integers, the largest of which is N, is precisely

$$N(N-1)\ldots(N-n+1) = {}_NP_n.$$

Furthermore, by equation (3-1-1),

$${}_NP_n = {}_nP_n \cdot \binom{N}{n} = n!\binom{N}{n}.$$

Since $\binom{N}{n}$ is an integer, it follows that $n! \mid {}_NP_n$, where, of course, $n!$ is the product of the first n positive integers. $\blacksquare$

The combinatorial technique we've used here is very powerful. To show that $n! \mid N(N-1)\ldots(N-n+1)$, we established a combinatorial interpretation of $N(N-1)\ldots(N-n+1)/n!$ $\left(\text{that is, } \binom{N}{n}\right)$; since $\binom{N}{n}$ is an integer, we were able to deduce the desired result. In the next two sections we shall further illustrate this technique by using it in the proofs of two additional divisibility theorems.

EXERCISES

1. Prove that a set S of n elements has precisely 2^n subsets (the empty set and S itself are counted as subsets).

2. Use Exercise 1 to give a combinatorial proof of the formula

$$2^n = 1 + \binom{n}{1} + \binom{n}{2} + \binom{n}{3} + \ldots + \binom{n}{n}.$$

3. Using the definition of $\binom{n}{r}$ as the number of r-combinations of a set S of n elements, prove combinatorially that

$$\binom{n}{r} = \binom{n-1}{r} + \binom{n-1}{r-1}.$$

4. Give a new proof of Theorem 3–2, using the principle of mathematical induction in conjunction with Exercise 3.

5. Prove that

$$\binom{r}{r} + \binom{r+1}{r} + \binom{r+2}{r} + \ldots + \binom{r+m}{r} = \binom{r+m+1}{r+1}.$$

[Hint: Use the principle of mathematical induction and the equation appearing in Exercise 3.]

6. Prove that

$$1 - \binom{n}{1} + \binom{n}{2} - \binom{n}{3} + \ldots + (-1)^r \binom{n}{r} = (-1)^r \binom{n-1}{r},$$

if $0 < r \leq n$. [Hint: Use the principle of mathematical induction and the equation appearing in Exercise 3.]

7. Prove that

$$1 + \binom{n}{2} + \binom{n}{4} + \binom{n}{6} + \ldots = 2^{n-1},$$

and that

$$\binom{n}{1} + \binom{n}{3} + \binom{n}{5} + \binom{n}{7} + \ldots = 2^{n-1}.$$

8. Prove that

$$\binom{n}{r} = \binom{n}{n-r}.$$

9. Prove that if p is a prime and $0 < a < p$, then $p \mid \binom{p}{a}$.

10. Prove the binomial theorem:

$$(x+y)^n = x^n + \binom{n}{1}x^{n-1}y + \binom{n}{2}x^{n-2}y^2 + \dots + \binom{n}{n-1}xy^{n-1} + y^n.$$

[Hint: Use the principle of mathematical induction and the equation appearing in Exercise 3.]

11. Prove that if p is a prime and n is a positive integer, then $p \mid n^p - n$. [Hint: Use the mathematical induction on n, Exercise 9, and the binomial theorem with $x=k$ and $y=1$.]

12. Suppose that the sequences of real numbers $a_0, a_1, a_2, a_3, \dots$ and $b_0, b_1, b_2, b_3, \dots$ satisfy the relation

$$a_n = \sum_{r=0}^{n} \binom{n}{r} b_r.$$

Then prove that

$$(-1)^n b_n = \sum_{s=0}^{n} \binom{n}{s} (-1)^s a_s.$$

[Hint: Substitute the formula for a_s into $\sum_{s=0}^{n} \binom{n}{s}(-1)^s a_s$.]

13. Recall the definition of the Fibonacci numbers: $F_1 = F_2 = 1$, and $F_n = F_{n-1} + F_{n-2}$ if $n > 2$. Prove that

$$F_n = 1 + \binom{n-2}{1} + \binom{n-3}{2} + \dots + \binom{n-j}{j-1} + \binom{n-j-1}{j},$$

where j is the largest integer less than or equal to $(n-1)/2$.

14. Prove that

$$\binom{n}{1}F_1 + \binom{n}{2}F_2 + \binom{n}{3}F_3 + \dots + \binom{n}{n-1}F_{n-1} + F_n = F_{2n},$$

where F_n denotes the nth Fibonacci number.

3–2 FERMAT'S LITTLE THEOREM

We shall use the combinatorial techniques of Section 3–1 to prove a result obtained by one of the founders of modern number theory, P. Fermat (1601–1665).

THEOREM 3–4: *If p is a prime and n is a positive integer, then* $p \mid n^p - n$.

PROOF: Suppose that we wish to form strings of p colored beads each, and that we have on hand enough beads to permit unlimited use of each of n colors. How many different strings can we form? By the general combinatorial principle, the answer is clearly n^p, since each bead may be chosen in exactly n ways and since there are p choices for each string. Figure 3–1 illustrates the case in which $n=3$ and $p=3$.

Of the n^p possibilities, exactly n strings have beads of only one color. Put these aside, and, in the manner illustrated in Figure 3–2, join the two ends of each of the remaining $n^p - n$ strings to form equally many bracelets.

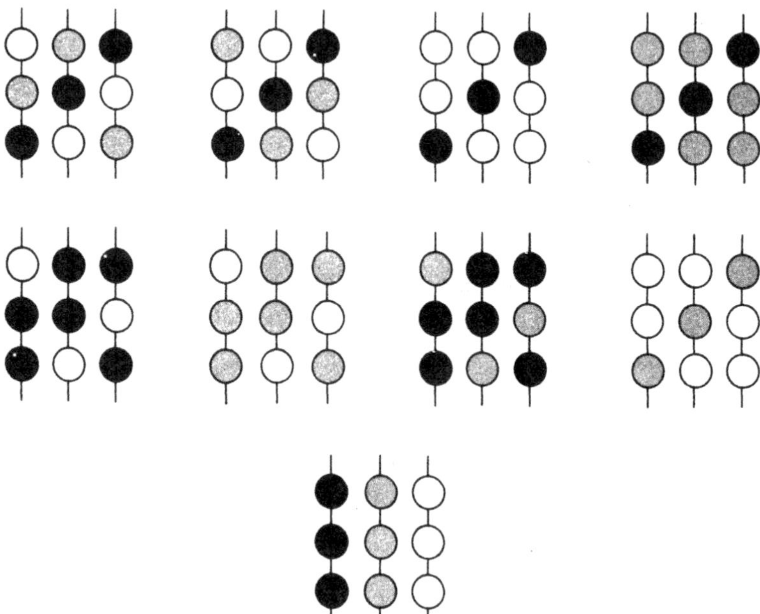

Figure 3–1 The twenty-seven strings of three beads with three possible colors.

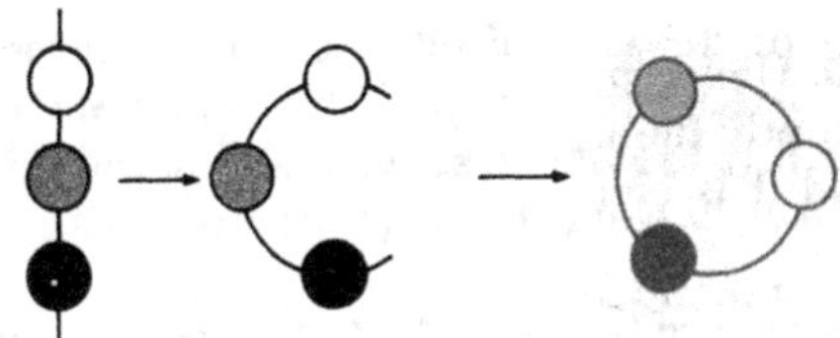

Figure 3–2 The formation of a bracelet from a string.

We can alter any string of beads by removing a bead from the top end and placing it on the bottom end; this alteration produces a different string of beads without changing the resulting bracelet. (When $n=3$ and $p=3$, the 24 possible multicolored strings fall naturally into eight groups of three strings obtainable from each other by one or more repetitions of this alteration [see the first eight groups in Figure 3–2]; to each of these eight groups there corresponds a distinct bracelet [see Figure 3–3].) Let k be the least number of times this alteration may be repeated before the original color scheme of the string is reproduced. Certainly $k > 1$, since we have set aside all monocolor strings. Note that after $2k$ steps the bracelet's original color scheme will again be reproduced; and, similarly, after $3k$ steps, $4k$ steps, and so on. By Euclid's division lemma (Theorem 2–1), there exist h and r such that

$$p = hk + r \qquad (0 \leq r < k).$$

Since the color scheme is reproduced after hk steps and is also reproduced after p steps (because after p steps all the beads are back in their original positions), it takes exactly r steps after the hkth step to get a reproduction of the color scheme. Since $r< k$ and k is the least

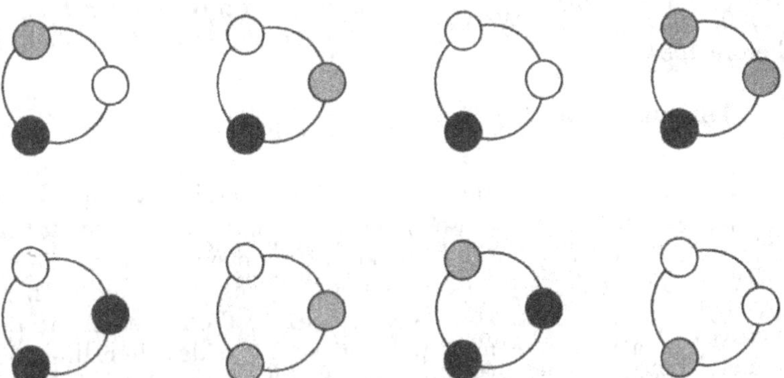

Figure 3–3 The eight bracelets of three beads with three possible colors (monocolor bracelets excluded).

positive number of steps required to obtain a reproduction, we see that r must be 0. Hence $p = hk$, and so $k = p$, since $k > 1$ and p is a prime. Consequently the $n^p - n$ strings fall naturally into groups of p strings each, and it is clear that each group gives rise to a separate bracelet.

Thus the number of bracelets N multiplied by p gives the number of strings that are not monocolor, namely $n^p - n$. Hence $pN = n^p - n$; that is, $p \mid n^p - n$. ∎

EXERCISES

1. Prove that if g.c.d.$(n,p) = 1$, then $p \mid n^{p-1} - 1$.

2. Prove that $12 \mid n^2 - 1$ if g.c.d. $(n,6) = 1$.

3. Prove that n^5 and n both have the same last digit.

4. In view of the fact that there are $n!$ n-permutations of n colors, does the proof of Theorem 3–4 provide a proof that $n! \mid n^p - n$?

5. Prove that if $3 \nmid n$, then $3 \mid n^2 - 1$.

6. Prove that if $5 \nmid (n-1)$, $5 \nmid n$, and $5 \nmid (n+1)$, then $5 \mid (n^2+1)$. [Hint: Multiply together n, $n-1$, $n+1$, and n^2+1.]

3–3 WILSON'S THEOREM

The theorem proved in this section was attributed to Sir John Wilson (1741–1793) by E. Waring in *Meditationes Algebraicae* (1770); however, G. W. Leibnitz appears to have discovered it before 1683.

Theorem 3–5: *If p is a prime, then $p \mid [(p-1)!+1]$.*

Proof: If $p = 2$, the theorem is obvious; therefore we can assume that p is an odd prime. Consider p points on a circle distributed so that they divide the circle into p equal arcs. How many polygons can we form by joining these points (crossing of edges is allowed)? These polygons are called stellated p-gons, because their vertices are the vertices of a regular convex polygon with p sides. Recalling the general combinatorial principle, one might expect ${}_pP_p(=p!)$ ways, since we may choose the first vertex in p ways, the second in $(p-1)$ ways, and so forth; however, note that we can describe each of the

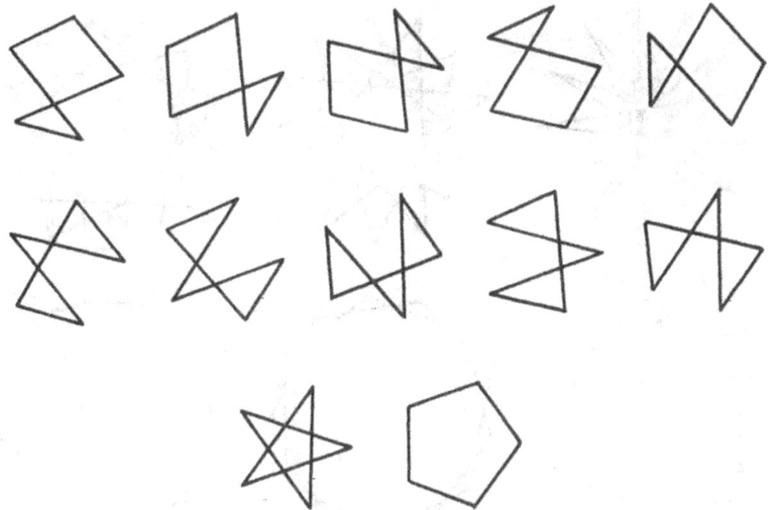

Figure 3–4 The twelve stellated pentagons.

p-gons in $2p$ different ways, namely, by starting at any one of its p vertices and choosing one or the other of the two segments at that vertex as the initial segment. Therefore, we really obtain $p!/2p$ different p-gons. Figure 3–4 shows the twelve stellated pentagons.

Of the $p!/2p$ p-gons, exactly $(p-1)/2$ are unaltered when rotated through an angle of $2\pi/p$ radians; such unalterable p-gons are called regular stellated p-gons since they are "stars" of p points, with each point the vertex of a $(2k+1)\pi/p$ angle $(0 \leq k < (p-1)/2)$. In the case $p=5$, there are two such pentagons, shown in the third row of Figure 3–4; in the case $p=13$, there are six unaltered 13-gons (triskaidecagons), as illustrated in Figure 3–5. The remaining $p!/2p-(p-1)/2$ stellated p-gons fall naturally into sets of p elements; the members of each set can be obtained from a single member by successive rotations through $2\pi/p$. The observation that there are p elements in each set may be verified as in the proof of Fermat's little theorem, where we showed that each bracelet arose from p strings of beads. When $p=5$, there are two such sets (they constitute the first and second rows of Figure 3–4).

Thus, the total number of sets is

$$\frac{\dfrac{p!}{2p}-\dfrac{p-1}{2}}{p}=\frac{(p-1)!-(p-1)}{2p}.$$

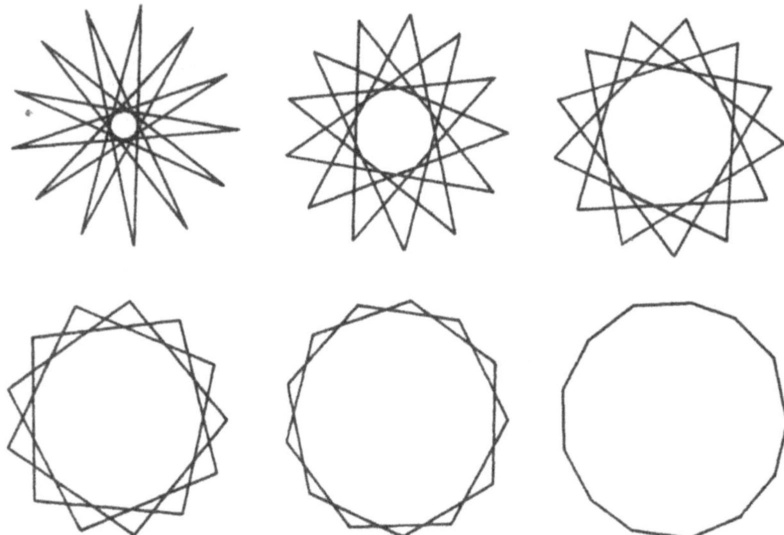

Figure 3-5 The six regular stellated 13-gons.

Hence $2p \mid [(p-1)!-p+1]$; consequently, $p \mid (p-1)!+1$, as desired. ■

EXERCISES

1. Prove that p is the smallest prime that divides $(p-1)!+1$.

2. Prove that $10 \nmid [(n-1)!+1]$, for each n.

3-4 GENERATING FUNCTIONS

This section is devoted to a technique with numerous applications throughout combinatorial analysis and the theory of numbers.

Definition 3-3: *If* $A = \{a_n\}_{n=b}^{\infty} = \{a_b, a_{b+1}, a_{b+2}, \ldots\}$ *is a sequence of real numbers, then*

$$f_A(x) = \sum_{n=b}^{\infty} a_n x^n$$

is called the **generating function** *for the sequence* A.

The function $f_A(x)$ is defined for all values of x for which the series converges. These admissible values of x will vary from sequence to sequence. In most cases the lower index b will be either 0 or 1.

Example 3–6: If $a_n = 1$ for all $n \geq 0$, then by the formula for the sum of an infinite geometric series,

$$f_A(x) = \sum_{n=0}^{\infty} x^n = \frac{1}{1-x}, \qquad |x| < 1.$$

(The second equality above follows if we let $n \to \infty$ in Theorem 1–2.)

Example 3–7: If $a_n = 1/n!$ for all $n \geq 0$, then

$$f_A(x) = \sum_{n=0}^{\infty} x^n/n! = e^x;$$

this series converges for all real values of x.

To illustrate the use of generating functions, we shall solve the following rather difficult problem:

Let a_n denote the number of ways of associating multiplicands with parentheses in an (ordered) product of n numbers so that the resulting expression correctly defines a multiplication of the numbers. What is a simple expression for a_n?

Now $a_1 = a_2 = 1$. Since three numbers may be associated in two ways, namely, $b_1(b_2b_3)$ and $(b_1b_2)b_3$, we see that $a_3 = 2$. Furthermore, $a_4 = 5$, since four numbers may be associated as follows: $((b_1b_2)b_3)b_4$, $(b_1(b_2b_3))b_4$, $(b_1b_2)(b_3b_4)$, $b_1(b_2(b_3b_4))$, $b_1((b_2b_3)b_4)$.

The following relationship holds among the a_i:

$$a_{n+1} = a_1a_n + a_2a_{n-1} + a_3a_{n-2} + \ldots + a_na_1. \tag{3-4-1}$$

The ith term in this sum, a_ia_{n-i+1}, is the number of associations of $n + 1$ terms in which the two outermost bracketings contain i terms and $n - i + 1$ terms, respectively; there are a_i ways of associating the terms inside the first set of brackets, and a_{n-i+1} ways of associating those inside the second set. The total number of bracketings is the sum of the number of bracketings with i terms in the first set of outermost brackets for $1 \leq i \leq n$.

Our objective now is to prove the following result.

Theorem 3–6: $a_n = \binom{2n-2}{n-1} \Big/ n.$

PROOF: We start with the defining formula

$$f_A(x) = \sum_{n=1}^{\infty} a_n x^n.$$

Then, by relation (3–4–1),

$$\begin{aligned} f_A(x) &= x + \sum_{n=2}^{\infty} (a_1 a_{n-1} + \ldots + a_{n-1} a_1) x^n \\ &= x + \left(\sum_{n=1}^{\infty} a_n x^n\right)\left(\sum_{n=1}^{\infty} a_n x^n\right) \\ &= x + (f_A(x))^2 \end{aligned}$$

(the penultimate equation follows directly from the formula for the multiplication of power series). Thus $f_A(x)$ satisfies the quadratic equation $y^2 - y + x = 0$.

Hence $f_A(x)$ is one of the two functions $\frac{1}{2} + \frac{(1-4x)^{1/2}}{2}$ and $\frac{1}{2} - \frac{(1-4x)^{1/2}}{2}$. Since $f_A(0) = 0$, we see that $f_A(x)$ is in fact the second of these. Thus

$$f_A(x) = \frac{1}{2} - \frac{1}{2}(1-4x)^{1/2}. \tag{3-4-2}$$

We now expand $(1-4x)^{1/2}$ into a binomial series:

$$f_A(x) = \frac{1}{2} - \frac{1}{2} \sum_{n=0}^{\infty} (-1)^n \binom{\frac{1}{2}}{n} 4^n x^n,$$

where

$$\binom{\frac{1}{2}}{0} = 1 \quad \text{and} \quad \binom{\frac{1}{2}}{n} = \frac{\frac{1}{2}\left(\frac{1}{2}-1\right)\ldots\left(\frac{1}{2}-n+1\right)}{n!}.$$

Consequently,

$$f_A(x) = -\frac{1}{2} \sum_{n=1}^{\infty} (-1)^n \binom{\frac{1}{2}}{n} 4^n x^n. \tag{3-4-3}$$

In passing we note that the series for $f_A(x)$ converges for $|x| < 1/4$; this may be proved by means of the ratio test.

Since a function has at most one MacLaurin series expansion, the coefficients of the series in equation (3–4–3) must be a_n. (Recall that a MacLaurin series expansion is a Taylor series expansion about the origin $x=0$.) Hence

$$\begin{aligned}
a_n &= -\frac{1}{2}(-1)^n\binom{\frac{1}{2}}{n}4^n \\
&= (-1)^{n-1}\frac{1}{2}\,\frac{\frac{1}{2}\left(\frac{1}{2}-1\right)\cdots\left(\frac{1}{2}-n+1\right)}{n!}4^n \\
&= (-1)^n\frac{1}{2}\,\frac{1\cdot(-1)(1-2)\cdots(1-2(n-1))2^n}{n!} \\
&= \frac{1}{2}\,\frac{1\cdot 3\cdots(2n-3)2^n}{n!} \\
&= \frac{1}{2}\,\frac{1\cdot 3\cdots(2n-3)\cdot n!\,2^n}{n!\cdot n!} \\
&= \frac{1}{2}\,\frac{1\cdot 2\cdot 3\cdot 4\cdots(2n-4)(2n-3)(2n-2)2n}{(n!)^2} \\
&= \frac{1}{n}\,\frac{1\cdot 2\cdot 3\cdot 4\cdots(2n-2)}{[(n-1)!]^2} \\
&= \binom{2n-2}{n-1}\Big/ n.
\end{aligned}$$

■

If you attempted to guess the form of a_n before examining Theorem 3–6, you undoubtedly had difficulty. Even if you had discovered the correct form, you would still have been faced with the arduous task of proving that it was the right form. However, with the aid of generating functions, we were able to establish easily the formula in Theorem 3–6.

EXERCISES

1. Prove that $x/(1-x)^2$ is the generating function for the sequence $a_1=1,\ a_2=2,\ \ldots,\ a_n=n,\ \ldots.$

2. Prove that $-\log(1-x)$ is the generating function for the sequence $a_1=1,\ a_2=1/2,\ \ldots,\ a_n=1/n,\ \ldots.$

3. Let $d_n=\binom{N}{n}\binom{M}{0}+\binom{N}{n-1}\binom{M}{1}+\binom{N}{n-2}\binom{M}{2}+\cdots+$

$\binom{N}{0}\binom{M}{n}$. Using the binomial theorem, prove that $(1+x)^N(1+x)^M$ is the generating function for d_n. Conclude that $d_n = \binom{N+M}{n}$.

4. Prove that $1/(1-x)^3$ is the generating function for the sequence of triangular numbers $a_0 = 1$, $a_1 = 3$, . . ., $a_n = \binom{n+2}{2}$, ...

5. Prove that $x/(1-x-x^2)$ is the generating function for the sequence of Fibonacci numbers $F_1 = 1$, $F_2 = 1$, . . ., $F_n = F_{n-1} + F_{n-2}$, ... [Hint: Prove that if $f_F(x) = \sum_{n=1}^{\infty} F_n x^n$, then $(1-x-x^2)f_F(x) = x$.]

6. Prove that $1/(1-2x)$ is the generating function for the sequence $a_0 = 1$, $a_1 = 2$, . . ., $a_n = 2^n$, ...

7. Prove that $\dfrac{x}{(1-bx)(1-bx-x)}$ is the generating function for the sequence $a_1 = 1$, $a_2 = 2b+1$, . . ., $a_n = (b+1)^n - b^n$, ...

8. What is the generating function for the sequence defined by $a_0 = 1$ and $a_n = 2$ for $n > 0$?

3-5 THE USE OF COMPUTERS IN NUMBER THEORY

This section is not intended as an introduction to programming. Rather, it is intended to suggest the importance of computation in number theory.

The computer has been used to prove a variety of theorems in number theory; for example, Donald Gillies used two hours and fifteen minutes on the Illiac II at the University of Illinois to establish that $2^{11213}-1$, a number of 3376 digits, is a prime. The University of Illinois was proud enough of this achievement to put "$2^{11213}-1$ IS PRIME" on its postage meter, so that it appeared on all outgoing mail. However, the computer's most important role has been to provide numerical data from which one may guess the truth of a theorem.

At the end of this section are a number of projects requiring only a modest knowledge of computer programming. Indeed, many of them are so simple that an energetic student with a desk calculator (or with several sharp pencils and a pot of coffee) may complete them.

To illustrate a typical project, along with conclusions that might be drawn, let us work through the following problem:

A perfect number is a number equal to the sum of its proper divisors (that is, all divisors except the number itself); find all perfect numbers $n (2 \leq n \leq 500)$.

To proceed, we need only write a program that determines for each n all proper divisors $d_1, d_2, \ldots, d_r$ of n, and then prints out n whenever $\sum_{i=1}^{r} d_i = n$.

The diagram in Figure 3–6 indicates the steps a computer could follow in finding all perfect numbers not exceeding 500; such a diagram is called a *flow diagram* or *flow chart.*

One may instead use a reference such as the *C.R.C. Standard Mathematical Tables* and a desk calculator to determine quickly the sum of the proper divisors of n. The perfect numbers not exceeding 500 are 6, 28, and 496.

It is easy to formulate conjectures about these numbers, even from these meager data. Note that $6 = 2 \cdot 3, 28 = 2^2 \cdot 7, 496 = 2^4 \cdot 31$. Thus, in each instance, the perfect number is the product of a power of 2 and a single odd prime; in particular, each perfect number not exceeding 500 is of the form $2^{p-1}(2^p - 1)$, where $2^p - 1$ is an odd prime.

CONJECTURE 1: $2^{p-1}(2^p - 1)$ is a perfect number whenever $2^p - 1$ is a prime.

CONJECTURE 2: All perfect numbers are of the form given in Conjecture 1.

We can easily prove Conjecture 1. Suppose $2^p - 1$ is a prime. Then, the proper divisors of 2^{p-1} $(2^p - 1)$ are $1, 2, 4, \ldots, 2^{p-1}, 2^p - 1, 2(2^p - 1), \ldots,$ and $2^{p-2}(2^p - 1)$. The sum of these numbers is

$$\sum_{j=0}^{p-1} 2^j + \sum_{j=0}^{p-2} (2^p - 1)2^j = 2^p - 1 + (2^p - 1)(2^{p-1} - 1) = 2^{p-1}(2^p - 1).$$

Conjecture 2 was raised by the ancient Greeks, and it is still not completely solved. L. Euler proved that every *even* perfect number is of the form given in Conjecture 1; however, it is not known whether or not any odd perfect numbers exist. We know only that none exist that are less than 10^{36} (this result was proven in 1967).

Any student of number theory who wishes to obtain an adequate knowledge of computer programming should consult one of the

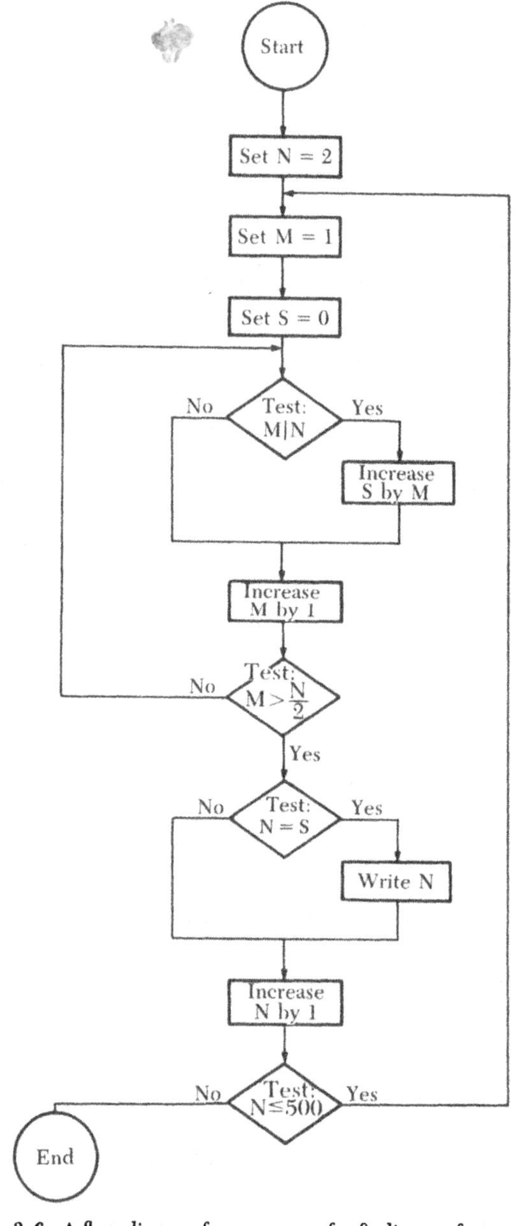

Figure 3–6 A flow diagram for a program for finding perfect numbers.

many excellent elementary texts, for example, *Basic Fortran IV Programming* by Healy and Debruzzi (a programmed self-instructional text), *Introduction to Fortran IV Programming* by Blatt, or *A Guide to Fortran Programming* by McCracken.

EXERCISES

The part of the text to which each computer program is relevant is listed at the end of the corresponding exercise. You are encouraged to form conjectures from the data you obtain with your programs.

1. Find the number $d(n)$ of divisors of n for $1 \leq n \leq 200$. (Chapters 2, 6, and 15)

2. Find all the primes smaller than 1000. (Chapters 2 and 8)

3. Find all integers smaller than 1000 that are either perfect squares or sums of two perfect squares. (Chapter 11)

4. Find all integers smaller than 1000 that are either perfect squares or sums of two or three perfect squares. (Chapter 11)

5. Find all integers smaller than 1000 that are either perfect squares or sums of two, three, or four perfect squares. (Chapter 11)

6. A number n is called a *quadratic residue modulo p* (where p is a prime) if $p \nmid n$, and if there exists an integer $m (1 \leq m < p)$ such that $p \mid m^2 - n$. For $p = 11$, find all numbers $r (1 \leq r < p)$ that are quadratic residues modulo p. Do the same for $p = 3, 5, 7, 17, 29$, and 61. (Chapter 9)

7. Using the results of Exercise 6, find the number of consecutive pairs of quadratic residues modulo p for $p = 3, 5, 7, 11, 17, 29$, and 61. Compare your answer with $p/4$ in each case. (Chapter 10)

8. Find the number of ways an integer n can be represented as a sum of distinct positive integers for $n \leq 30$ (for example, 5 can be represented in three ways: 5, $4 + 1$, and $3 + 2$). (Chapter 12)

9. Find the number of ways an integer n ($n \leq 30$) can be represented as a sum of odd positive integers (for example, 5 can be represented in three ways: 5, $3+1+1$, and $1+1+1+1+1$). Compare your results with those obtained for Exercise 8. (Chapter 12)

10. Find the sum $\sigma(n)$ of the divisors of n for $1 \leq n \leq 200$. (Chapter 6)

11. Let $\phi(n)$ denote the number of positive integers not exceeding n that are relatively prime to n; find $\phi(n)$ for $1 \leq n \leq 100$. (Chapter 6)

12. Let $b(n)$ denote the number of ways of representing n as a sum of nonnegative powers of 2 (for example, $b(4)=4$, since 4 has the representations 4, $2+2$, $2+1+1$, $1+1+1+1$); find $b(n)$ for $1 \leq n \leq 40$. Are there any significant relationships between $b(2n+1)$, $b(2n)$, $b(n)$, and $b(n-1)$? (Chapter 13)

13. For $1 \leq n \leq 15$, compute the number $t(n)$ of representations of n as a sum of positive integers that are not multiples of 3 (for example, $t(4)=4$, since 4 has the representations $4, 2+2, 2+1+1, 1+1+1+1$). (Chapter 12)

14. For $1 \leq n \leq 15$, compute the number $v(n)$ of representations of n as a sum of positive integers in which no summand appears more than twice (for example, $v(4)=4$, since 4 has the representations $4, 3+1, 2+2, 2+1+1$). Compare $v(n)$ with $t(n)$. (Chapter 12)

CHAPTER 4

FUNDAMENTALS OF CONGRUENCES

In Chapter 3, we selected the proofs of Fermat's little theorem and Wilson's theorem for illumination rather than for brevity. More succinct proofs of these theorems are possible with congruence notation, first introduced by Gauss. In this chapter and in Chapter 5, we shall investigate the many important properties of congruences.

4–1 BASIC PROPERTIES OF CONGRUENCES

DEFINITION 4–1: If $c \neq 0$, we say that $a \equiv b \pmod{c}$ (read "a is congruent to b modulo c") provided that $(a-b)/c$ is an integer (that is, provided that $c \mid a-b$).

Example 4–1: Since $(8-2)/2=3$ is an integer, we see that $8 \equiv 2 \pmod{2}$; but $8 \not\equiv 3 \pmod{2}$, because $(8-3)/2=2.5$ is not an integer. Similarly, $17 \equiv 12 \pmod{5}$, $100 \equiv -40 \pmod{20}$, and $11 \equiv -1 \pmod{12}$.

Example 4–2: We can now express Fermat's little theorem in congruence notation:

$$n^p \equiv n \pmod{p};$$

similarly, we can state Wilson's theorem in the form

$$(p-1)! \equiv -1 \pmod{p}.$$

Example 4–3: If $c \neq 0$ and $a=b$, then $a \equiv b \pmod{c}$. Thus, equal integers are congruent modulo any nonzero integer; however, congruent integers are not necessarily equal.

If $(a-b)/c$ is an integer, so is $(a-b)/(-c)=-(a-b)/c$; hence $a \equiv b \pmod{c}$ if and only if $a \equiv b \pmod{-c}$. Thus congruences with negative moduli may be replaced by equivalent congruences with positive moduli. Throughout this chapter we shall assume that all moduli are positive.

The relation "is congruent to" is an equivalence relation. For those who are unfamiliar with equivalence relations, we shall now prove that the congruence relation has all the defining properties of an equivalence relation.

Theorem 4–1: *If a, b, c, and d are any integers* $(c \neq 0)$, *the following assertions hold:*

$$a \equiv a \pmod{c}; \tag{4-1-1}$$

$$\text{if } a \equiv b \pmod{c}, \text{ then } b \equiv a \pmod{c}; \text{ and} \tag{4-1-2}$$

$$\text{if } a \equiv b \pmod{c} \text{ and } b \equiv d \pmod{c}, \text{ then } a \equiv d \pmod{c}. \tag{4-1-3}$$

Remark: Statements (4–1–1), (4–1–2), and (4–1–3) are the *reflexive*, *symmetric*, and the *transitive* properties of congruences, respectively.

Proof: $(a-a)/c = 0$ is an integer; thus (4–1–1) holds. If $(a-b)/c$ is an integer, then $(b-a)/c = -(a-b)/c$ is also an integer; thus (4–1–2) is true. Finally, if $(a-b)/c$ and $(b-d)/c$ are integers, then so is $(a-d)/c$, because it is the sum of the first two. Thus (4–1–3) is confirmed. ■

The next theorem shows that congruences may be validly added, subtracted and multiplied.

Theorem 4–2: *Suppose* $a \equiv a' \pmod{c}$ *and* $b \equiv b' \pmod{c}$; *then*

$$a \pm b \equiv a' \pm b' \pmod{c}, \text{ and} \tag{4-1-4}$$

$$ab \equiv a'b' \pmod{c}. \tag{4-1-5}$$

Proof: Since $(a-a')/c$ and $(b-b')/c$ are integers, so are $\dfrac{(a \pm b) - (a' \pm b')}{c} = \dfrac{a-a'}{c} \pm \dfrac{(b-b')}{c}$ and $\dfrac{ab - a'b'}{c} = a\dfrac{b-b'}{c} + b'\dfrac{a-a'}{c}$. ■

Example 4–4: Since

$$19 \equiv 11 \pmod{4} \tag{4-1-6}$$

and

$$6 \equiv 2 \pmod 4, \tag{4-1-7}$$

we see that

$$\begin{aligned} 25 &\equiv 13 \pmod 4 && \text{(by addition of (4–1–6) and (4–1–7))}, \\ 13 &\equiv 9 \pmod 4 && \text{(by subtraction of (4–1–7) from (4–1–6))}, \\ 114 &\equiv 22 \pmod 4 && \text{(by multiplication of (4–1–6) and (4–1–7))}. \end{aligned}$$

Note that addition, subtraction, and multiplication of two congruences are permissible only when the congruences have the same moduli.

Division of both sides of a congruence by an integer is not always permissible; for example, in the congruence $14 \equiv 4 \pmod{10}$, it is not valid to divide both sides by 2, because $7 \not\equiv 2 \pmod{10}$. The following theorem provides conditions under which division *is* permissible.

THEOREM 4–3 *(Cancellation Law): If $bd \equiv bd' \pmod c$ and if g.c.d.$(b,c) = 1$, then $d \equiv d' \pmod c$.*

PROOF: Since $(bd - bd')/c$ is an integer, $c \mid b(d - d')$. Thus, by Theorem 2–3, $c \mid d - d'$, that is, $d \equiv d' \pmod c$. ■

Example 4–5: Since $6 \equiv 12 \pmod 2$ and g.c.d.$(2,3) = 1$, we can conclude that $2 \equiv 4 \pmod 2$; however, we cannot conclude that $3 \equiv 6 \pmod 2$, since g.c.d.$(2,2) = 2 \neq 1$.

We have now established procedures for performing algebraic operations on congruences much like the customary operations on ordinary equations.

EXERCISES

1. Find integers x such that
   (a) $5x \equiv 4 \pmod 3$
   (b) $7x \equiv 6 \pmod 5$
   (c) $9x \equiv 8 \pmod 7$.

2. Do there exist integers x such that
   (a) $6x \equiv 5 \pmod 4$
   (b) $10x \equiv 8 \pmod 6$
   (c) $12x \equiv 9 \pmod 6$?

3. Prove that if $x \equiv y \pmod{m}$ and $a_0, a_1, \ldots, a_r$ are integers, then $a_0x^r + a_1x^{r-1} + \ldots + a_r \equiv a_0y^r + a_1y^{r-1} + \ldots + a_r \pmod{m}$.

4. Prove that if $bd \equiv bd' \pmod{p}$, where p is a prime and $p \nmid b$, then $d \equiv d' \pmod{p}$.

5. Prove that if $|a| < k/2$, $|b| < k/2$, and $a \equiv b \pmod{k}$, then $a = b$.

6. Is there a perfect square n^2 such that

$$n^2 \equiv -1 \pmod{p} \qquad (4\text{-}1\text{-}8)$$

for $p=3$? for $p=5$? for $p=7$? for $p=11$? for $p=13$? for $p=17$? for $p=19$? Can you characterize the primes p for which congruence (4–1–8) has a solution?

7. Which of the following congruences hold?

(a) $12{,}345{,}678{,}987{,}654{,}321 \equiv 0 \pmod{12{,}345{,}678}$

(b) $12{,}345{,}678{,}987{,}654{,}321 \equiv 0 \pmod{12{,}345{,}679}$

(c) $57 \equiv 208 \pmod{4}$

(d) $531 \equiv 1236 \pmod{7561}$

(e) $12321 \equiv 111 \pmod{3}$.

4–2 RESIDUE SYSTEMS

Theorem 4–2 assures us that, in a congruence involving only addition, subtraction, and multiplication, we may validly replace integers with congruent integers. We shall now study some important concepts that will facilitate making these replacements.

DEFINITION 4–2: *If h and j are two integers and $h \equiv j \pmod{m}$, then we say that j is a* **residue** *of h* modulo m.

DEFINITION 4–3: *The set of integers $\{r_1, r_2, \ldots, r_s\}$ is called a* **complete residue system** modulo m if

(i) *$r_i \not\equiv r_j \pmod{m}$ whenever $i \neq j$;*

(ii) *for each integer n there corresponds an r_i such that $n \equiv r_i \pmod{m}$.*

Theorem 4–4: *If s different integers $r_1, r_2, \ldots, r_s$ form a complete residue system* modulo *m, then $s = m$.*

Proof: Let $t_0 = 0$, $t_1 = 1$, $t_2 = 2, \ldots,$ and $t_{m-1} = m - 1$. Then the m integers $t_0, t_1, \ldots, t_{m-1}$ form a complete residue system modulo m. To see this, note that for each n there exist unique integers q and u such that

$$n = mq + u \qquad \text{and} \qquad 0 \leq u < m$$

(by Euclid's division lemma). Hence $n \equiv u \pmod{m}$, and u is one of the t_i. Furthermore, since $|t_i - t_j| \leq m - 1$, no two of the t_i are congruent modulo m. Consequently, we see that the set $\{t_0, t_1, \ldots, t_{m-1}\}$ is a complete residue system modulo m.

Thus, each r_i is congruent to exactly one of the t_i (so that $s \leq m$). Conversely, since the r_i form a complete residue system, every t_i is congruent to exactly one of the r_i (so that $m \leq s$). Hence $s = m$. ■

Corollary 4–1: *Let m be a positive integer; then $\{0,1,2, \ldots, m-1\}$ is a complete residue system* modulo *m.*

Example 4–6: The sets $\{1,2,3\}$, $\{0,1,2\}$, $\{-1,0,1\}$, and $\{1,5,9\}$ are all complete residue systems modulo 3.

When working with congruences modulo m, we can replace the integers in the congruences by elements of $\{0,1,2, \ldots, m-1\}$. Such substitutions often lead to the simplification of seemingly complicated problems.

Example 4–7: Find an integer n that satisfies the congruence

$$325n \equiv 11 \pmod{3}. \tag{4-2-1}$$

Since

$$325 \equiv 1 \pmod{3}$$

and

$$11 \equiv 2 \pmod{3},$$

we must find an integer n such that

$$n \equiv 2 \pmod{3}.$$

The answer is now obvious; the integer 2 will do quite nicely.

In our study of congruences modulo m, we shall often be interested in the integers that are relatively prime to m. The elements of a complete residue system that are relatively prime to m form what is called a reduced residue system modulo m.

DEFINITION 4–4: *The set of integers* $\{r_1,r_2,\ldots,r_s\}$ *is called a* **reduced residue system** modulo m if

(i) g.c.d.$(r_i,m)=1$ *for each* i;

(ii) $r_i \not\equiv r_j \pmod m$ *whenever* $i \neq j$; *and*

(iii) *for each integer* n *relatively prime to* m *there corresponds an* r_i *such that* $n \equiv r_i \pmod m$.

Example 4–8: The set $\{0,1,2,3,4,5\}$ is a complete residue system modulo 6; consequently $\{1,5\}$ is a reduced residue system modulo 6. In general, we can obtain a reduced residue system from a complete residue system by deleting those elements of the complete residue system that are not relatively prime to m.

Example 4–9: Let p denote a prime; then $\{0,1,2,\ldots,p-1\}$ is a complete residue system modulo p. The only element of this set not relatively prime to p is 0; hence $\{1,2,\ldots,p-1\}$ is a reduced residue system modulo p. In general, if p is a prime, we can obtain a reduced residue system modulo p by deleting the one element in a complete residue system that is divisible by p. Thus, we see that a reduced residue system modulo p has $p-1$ elements.

DEFINITION 4–5: *The function* $\phi(m)$ *shall denote the number of positive integers less than or equal to* m *that are relatively prime to* m. *This function* $\phi(m)$ *is called the Euler* ϕ*-function.*

THEOREM 4–5: *If* s *integers* $r_1,r_2,\ldots,r_s$ *form a reduced residue system* modulo m, *then* $s=\phi(m)$.

PROOF: This proof is similar to that of Theorem 4–4, except that now we let $t_1, t_2, \ldots, t_{\phi(m)}$ denote the $\phi(m)$ positive integers not exceeding m that are relatively prime to m. For each integer n relatively prime to m, Euclid's division lemma guarantees the existence of integers q and u such that,

$$n=qm+u \qquad \text{and} \qquad 0 \leq u < m.$$

If g.c.d.$(u,m)=d$, then $d \mid u$ and $d \mid m$; thus $d \mid n$, and so $d \mid$ g.c.d.(n,m). However, since n is relatively prime to m, g.c.d.$(n,m)=1$; consequently, $d=1$, and we see that u and m are relatively prime. Hence, u is one of the t_i. Since $|t_i-t_j| < m-1$, no two of the t_i are congruent modulo m; thus, the t_i form a reduced residue system modulo m.

Therefore, every r_i is congruent to exactly one t_i; thus $s \leq \phi(m)$. Conversely, since the r_i form a reduced residue system, each t_i is congruent to precisely one of the r_i; thus $\phi(m) \leq s$. Hence $s = \phi(m)$. ■

The study of residue systems belongs both to number theory and to algebra. Students with a knowledge of modern algebra will recognize that the complete residue system constitutes a finite ring [r_i "+" r_j is defined as the element r_h such that $r_h \equiv r_i + r_j \pmod{m}$; r_i "·" r_j is that r_h for which $r_h \equiv r_i r_j \pmod{m}$]. The reduced residue system is simply the group of units in this finite ring. Since we are concerned with the properties of integers rather than with the algebraic systems that may be constructed from them, we refer students interested in the algebraic approach to Chapter 1 of *A Survey of Modern Algebra* by G. Birkhoff and S. MacLane.

EXERCISES

1. Which of the following are complete residue systems modulo 11?

   (a) 0, 1, 2, 4, 8, 16, 32, 64, 128, 256, 512

   (b) 1, 3, 5, 7, 9, 11, 13, 15, 17, 19, 21

   (c) 2, 4, 6, 8, 10, 12, 14, 16, 18, 20, 22

   (d) −5, −4, −3, −2, −1, 0, 1, 2, 3, 4, 5.

2. Which of the following are reduced residue systems modulo 18?

   (a) 1, 5, 25, 125, 625, 3125

   (b) 5, 11, 17, 23, 29, 35

   (c) 1, 25, 49, 121, 169, 289

   (d) 1, 5, 7, 11, 13, 17.

3. Suppose $\{a_1, a_2, \ldots, a_k\}$ is a complete residue system modulo k, where k is a prime. Prove that for each integer n and each nonnegative integer s there exists a congruence of the form

$$n \equiv \sum_{j=0}^{s} b_j k^j \pmod{k^{s+1}}$$

   where each b_j is one of the a_i.

4. Let $w(n)$ denote the number of primes not exceeding n that do not divide n. Is $w(n) < \phi(n)$? Can you find values of n for which $w(n) = \phi(n) - 1$?

4-3 RIFFLING

In this section we shall demonstrate the usefulness of congruences in solving a problem of frequent interest to college students.

Take an ordinary deck of cards arranged in any order. How many shuffles are required before the deck returns to its original order? Of course, to idealize the real life situation, we must stipulate that only the modified perfect faro shuffle is permitted, as follows: Cut the deck into two equal 26-card packs; then proceed by alternating cards from each pack. The cards formerly in positions 1,2,...,26 are moved to positions 2,4,...,52; and the cards formerly in positions 27,28,...,52 are moved to positions 1,3,...,51. Thus, if a card starts in position x, it will end in position y, where $1 \leq y \leq 52$ and $2x \equiv y \pmod{53}$. After n shuffles, the card will be in position w, where $1 \leq w \leq 52$ and $2^n x \equiv w \pmod{53}$.

To determine the number of necessary shuffles, we must find n such that $2^n x \equiv x \pmod{53}$ for every x such that $1 \leq x \leq 52$. Since 53 is a prime, we may cancel x from both sides of the congruence; thus we must solve the congruence

$$2^n \equiv 1 \pmod{53}. \tag{4-3-1}$$

By Fermat's little theorem, we know that

$$2^{53} \equiv 2 \pmod{53}. \tag{4-3-2}$$

Cancelling 2 from both sides of congruence (4–3–2), we find that

$$2^{52} \equiv 1 \pmod{53}. \tag{4-3-3}$$

Hence, the cards will return to their original order after 52 shuffles. Actually, 52 is the least number of shuffles required, but we shall not prove this here.

In general, if we have a deck of m cards, then n shuffles will return the cards to the original order provided that

$$2^n \equiv 1 \pmod{m+1}. \tag{4-3-4}$$

Thus, if $m = 62$, we need only 6 shuffles, since $2^6 - 1 = 63$.

EXERCISES

1. How many modified perfect faro shuffles are needed to return the cards to their original position in a deck of 6 cards? of 8 cards? of 12 cards?

2. Suppose that instead of performing a modified perfect faro shuffle as described in this section, we shuffle as follows: Take the bottom and top cards of the deck and place them on the table to start a new deck. Then take the remaining bottom and top cards and place them on the newly started pile. Continue this process until all cards are gone from the original pack. For example, if the deck has six cards, then we shuffle as shown in Figure 4–1.

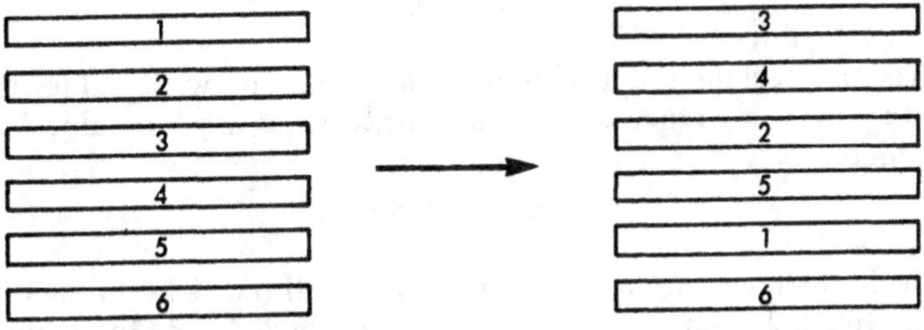

Figure 4–1 Shuffle of a deck of six cards described in Exercise 2 (side view).

Prove that the cards in a deck of 2^n cards will return to their original positions after $n+1$ such shuffles. (Note that the shuffle described in this problem is mechanically somewhat easier to perform than the modified perfect faro shuffle described earlier.)

CHAPTER 5

SOLVING CONGRUENCES

In Chapter 2, we studied the problem of finding integers x and y that satisfy the linear Diophantine equation

$$ax + by = c.$$

We can restate the problem as follows: For what values of x is $(ax - c)/b$ an integer? In other words, what are the values of x for which

$$ax \equiv c \pmod{b}?$$

In the first section of this chapter, we shall translate the theorem on the linear Diophantine equation (Theorem 2–4) into the language of congruences.

We are often interested in solving several congruences simultaneously. Such a problem arose relatively early in the history of mathematics. In the first century A.D., Sun-Tsu asked: What number yields the remainders 2, 3, and 2 when divided by 3, 5, and 7, respectively? In terms of congruences, Sun-Tsu was asking for an integer n that satisfies the three conditions

$$n \equiv 2 \pmod{3},$$

$$n \equiv 3 \pmod{5},$$

and

$$n \equiv 2 \pmod{7}.$$

In Section 5–3, we shall learn how to solve Sun-Tsu's problem.

5–1 LINEAR CONGRUENCES

The simplest congruence problem may be phrased as follows: If a, b, and c are integers, can we find an integer n such that

$$an \equiv b \pmod{c}? \qquad (5\text{-}1\text{-}1)$$

Note that if n does satisfy this congruence, then all integers of the form $n + kc$ are also solutions, because

$$a(n + kc) = an + akc \equiv an \equiv b \pmod{c}.$$

Example 5–1: If $a = 5$, $b = 3$, and $c = 8$, we see that 7 is a solution of the congruence

$$5n \equiv 3 \pmod{8}.$$

Indeed, each of the numbers $\ldots, -17, -9, -1, 7, 15, 23, 31, \ldots$ is a solution.

So we see that if we can find a solution n of congruence (5–1–1), we can then construct infinitely many others. However, all the solutions we have constructed so far are congruent to n modulo c. This leads us to ask: How many mutually incongruent solutions of congruence (5–1–1) are there?

To answer our two questions about congruence (5–1–1), we use the knowledge of the linear Diophantine equation we obtained in Theorem 2–4. Solving (5–1–1) is equivalent to finding integers n and k such that $an - b = kc$, that is,

$$an + (-c)k = b. \qquad (5\text{-}1\text{-}2)$$

Now, by Theorem 2–4, integers n and k satisfying equation (5–1–2) exist if and only if $d \mid b$, where $d = \text{g.c.d.}(a,c)$. *Thus* (5–1–1) *has a solution* n *if and only if* $d \mid b$, *where* $d = \text{g.c.d.}(a,c)$.

To answer our second question, we note that, by Theorem 2–4, each solution of (5–1–2) has the form

$$n = n_0 + ct/d \qquad k = k_0 + at/d, \qquad (5\text{-}1\text{-}3)$$

where n_0 and k_0 constitute one specific solution, t is any integer, and $d = \text{g.c.d.}(a,c)$. Among the different values of n described by (5–1–3), we note that the d values $n_0, n_0 + c/d, n_0 + 2c/d, \ldots, n_0 + (d-1)c/d$ are all mutually incongruent modulo c, because the absolute difference between any two of them is less than c. If $n = n_0 + ct/d$ is any other value, we write $t = qd + r$, where $0 \leq r < d$, and we see that

$$\begin{aligned} n &= n_0 + c(qd + r)/d \\ &= n_0 + cq + cr/d \\ &\equiv n_0 + cr/d \pmod{c}. \end{aligned}$$

Thus every solution of (5–1–2) (and therefore, of (5–1–1)) is congruent to one of the d values of n given above. *Hence there exist* $d = \text{g.c.d.}(a,c)$ *mutually incongruent solutions of (5–1–1), if there exist any.*

We now state our knowledge of the congruence (5–1–1) as a theorem.

THEOREM 5–1: *If* $d = \text{g.c.d.}(a,c)$, *then the congruence*

$$an \equiv b \pmod{c}$$

has no solution if $d \nmid b$, *and it has* d *mutually incongruent solutions if* $d \mid b$.

Example 5–2: Since $\text{g.c.d.}(5,8) = 1$, the congruence in Example 5–1 has a solution (as we already know), and all solutions of it are mutually congruent.

Example 5–3: Since $\text{g.c.d.}(21,3) = 3$ and $3 \nmid 11$, there exist no solutions of the congruence

$$21x \equiv 11 \pmod{3}.$$

Example 5–4: Since $\text{g.c.d.}(15,12) = 3$ and $3 \mid 9$, the congruence

$$15x \equiv 9 \pmod{12}$$

has exactly 3 mutually incongruent solutions. By inspection, we see that $x = 3$ is a solution. By Theorem 2–4, all solutions are therefore given by

$$x = 3 + t\frac{12}{3} = 3 + 4t \qquad t = \ldots, -2, -1, 0, 1, 2, \ldots.$$

By letting $t = 0,1,2$, we obtain 3 mutually incongruent solutions of the congruence: 3, 7, and 11.

We shall apply Theorem 5–1 to the particular case $b = 1$ after introducing two new terms.

DEFINITION 5–1: *We say that a solution* n *of a congruence (such as (5–1–1)) is* **unique** modulo c *if any solution* n' *of it is congruent to* n modulo c.

DEFINITION 5–2: *If* $a\tilde{a} \equiv 1 \pmod{c}$, *we say that* $\tilde{a}$ *is the* **inverse** *of* a modulo c.

COROLLARY 5–1: *If* $\text{g.c.d.}(a,c) = 1$, *then* a *has an inverse, and it is unique* modulo c.

PROOF: If g.c.d.$(a,c) = 1$, Theorem 5-1 implies that

$$an \equiv 1 \pmod{c}$$

has a solution n, and that it is unique modulo c. ■

Example 5-5: Since $5^2 = 25 \equiv 1 \pmod{8}$, we see that 5 is its own inverse modulo 8. We observe that -3 and 13 are also inverses of 5 modulo 8; this is not inconsistent with the fact that 5 has a unique inverse modulo 8, because $-3 \equiv 5 \equiv 13 \pmod{8}$.

EXERCISES

1. Find a complete set of mutually incongruent solutions of each of the following.

   (a) $7x \equiv 5 \pmod{11}$

   (b) $8x \equiv 10 \pmod{30}$

   (c) $9x \equiv 12 \pmod{15}$.

2. Which of the following congruences have solutions?

   (a) $99x \equiv 100 \pmod{101}$

   (b) $400898x \equiv 22 \pmod{400900}$

   (c) $27x \equiv 1 \pmod{51}$

   (d) $99x \equiv 100 \pmod{102}$

   (e) $30x \equiv 42 \pmod{49}$

   (f) $81x \equiv 57 \pmod{117}$.

3. Find $\tilde{a}$, the inverse of a modulo c, when

   (a) $a = 2$ and $c = 5$

   (b) $a = 7$ and $c = 9$

   (c) $a = 12$ and $c = 17$.

5-2 THE THEOREMS OF FERMAT AND WILSON REVISITED

Our knowledge of congruences leads to proofs of Fermat's little theorem (Theorem 3-4) and Wilson's theorem (Theorem 3-5) that are less picturesque than the proofs using a combinatorial approach, but that provide more general results.

Theorem 5–2 *(Euler's Theorem): If* g.c.d.$(a,m) = 1$, then

$$a^{\phi(m)} \equiv 1 \pmod{m}.$$

Proof: Let $r_1, r_2, \ldots, r_{\phi(m)}$ be a reduced residue system modulo m. We note that $ar_1, ar_2, \ldots, ar_{\phi(m)}$ are all relatively prime to m; furthermore, they are mutually incongruent, since $ar_i \equiv ar_j \pmod{m}$ implies that $r_i \equiv r_j \pmod{m}$, by the cancellation law (Theorem 4–3). We may thus pair each ar_i with some r_j such that $ar_i \equiv r_j \pmod{m}$, and we note that r_j is uniquely defined for each ar_i. Note also that each r_j is paired with some ar_i, since there are $\phi(m)$ of the r_j and $\phi(m)$ of the ar_i. Thus

$$r_1 r_2 \ldots r_{\phi(m)} \equiv ar_1 ar_2 \ldots ar_{\phi(m)} \pmod{m}.$$

Hence, if $R = r_1 r_2 \ldots r_{\phi(m)}$, then

$$R \equiv a^{\phi(m)} R \pmod{m}.$$

Now g.c.d.$(R,m) = 1$, because R is a product of integers each of which is relatively prime to m. Thus

$$a^{\phi(m)} \equiv 1 \pmod{m},$$

by the cancellation law. ■

Corollary 5–2 *(Fermat's Little Theorem, Theorem 3–4): If p is a prime, then*

$$n^p \equiv n \pmod{p}.$$

Proof: If $p \mid n$, then $n^p \equiv 0 \equiv n \pmod{p}$. If $p \nmid n$, then g.c.d.$(p,n) = 1$. Thus, by Theorem 5–2,

$$n^{p-1} \equiv 1 \pmod{p},$$

since $\phi(p) = p - 1$. Multiplying both sides of this congruence by n, we find that

$$n^p \equiv n \pmod{p}. \quad ■$$

We now prove a form of Wilson's theorem that is stronger than the one given in Chapter 3.

Theorem 5–3: *The congruence $(m-1)! \equiv -1 \pmod{m}$ holds if and only if m is a prime.*

PROOF: First suppose m is a prime, and consider the $m-1$ positive integers $1,2,\ldots,m-1$. Corollary 5-1 confirms that if a is such an integer, then there exists an inverse $\tilde{a}$ of a modulo m ($\tilde{a}$ is unique if we require that $1 \le \tilde{a} \le m-1$) for which

$$a\tilde{a} \equiv 1 \pmod{m}.$$

It may happen that a is its own inverse; in other words, that

$$a^2 \equiv 1 \pmod{m}.$$

In that case, $m \mid (a-1)(a+1)$; since m is prime, we see by Corollary 2-3 that either $m \mid a-1$ or $m \mid a+1$; therefore, $a \equiv \pm 1 \pmod{m}$. In the product $(m-2)(m-3)\ldots 2$, we pair each number with its inverse modulo p. Thus, for $m=7$, we write that $5\cdot 4\cdot 3\cdot 2=(5\cdot 3)\cdot(4\cdot 2)$. In general, we see that

$$\begin{aligned}(m-1)! &\equiv (m-1)\cdot 1\cdot\cdots\cdot 1 \pmod{m}\\ &\equiv m-1 \equiv -1 \pmod{m}.\end{aligned}$$

Conversely, suppose that m is not a prime. Then there exists an $a(1<a<m)$ such that $a \mid m$; note also that $a \mid (m-1)!$. If $(m-1)! \equiv -1 \pmod{m}$, then there exists an integer k such that $(m-1)!+1=km$. Since $a \mid m$ and $a \mid (m-1)!$, it follows from the last equation that $a \mid 1$; but this is an impossibility, because $a > 1$. Thus the congruence $(m-1)! \equiv -1 \pmod{m}$ cannot hold if m is not a prime. ■

EXERCISES

1. If $m=13$, then a reduced residue system modulo m is 1,2,3,4,5,6,7,8,9,10,11,12. Letting $a=3$, exhibit the pairing of each of the preceding numbers with the numbers in the reduced residue system 3,6,9,12,15,18,21,24,27,30, 33,36, as described in the proof of Theorem 5-2.

2. If $m=11$, then a reduced residue system modulo m is 1,2,3,4,5,6,7,8,9,10. Exhibit the pairing of each of the preceding numbers with its inverse modulo m as described in the proof of Theorem 5-3.

3. Prove that $1+a+a^2+\ldots+a^{\phi(m)-1} \equiv 0 \pmod{m}$ if g.c.d.$(a,m)=1$ and g.c.d.$(a-1,m)=1$.

4. Prove that if $r_1, r_2, \ldots, r_{\phi(m)}$ is a reduced residue system modulo m, and m is odd, then $r_1+r_2+\ldots+r_{\phi(m)} \equiv 0 \pmod{m}$.

5. What is the remainder when 41^{75} is divided by 3?

6. What is the remainder when 473^{38} is divided by 5?

7. Prove that if $A = a_0 10^n + a_1 10^{n-1} + \ldots + a_n$ and $S = a_0 + a_1 + \ldots + a_n$, then $A \equiv S \pmod 9$. (This result is the basis for the computational check called "casting out nines".)

8. Suppose that p is a prime, and that $p \nmid a$; use Euler's theorem to prove that $x = a^{p-2}b$ is a solution of

$$ax \equiv b \pmod p.$$

9. Prove that if p is a prime congruent to 1 modulo 4, then

$$\left(\frac{p-1}{2}\right)!^2 \equiv -1 \pmod p.$$

[Hint: Prove that $((p-1)/2)!^2 \equiv (p-1)! \pmod p$.]

10. Use Exercise 9 to find a solution of each of the following congruences.

(a) $x^2 \equiv -1 \pmod{13}$

(b) $x^2 \equiv -1 \pmod{17}$.

11. Prove that for each odd prime p and for each a $(0 \le a \le p-1)$,

$$\binom{p-1}{a} \equiv (-1)^a \pmod p.$$

12. Prove that for each prime p $(n < p \le 2n)$,

$$\binom{2n}{n} \equiv 0 \pmod p,$$

but that

$$\binom{2n}{n} \not\equiv 0 \pmod{p^2}.$$

13. Is $2^{p-1} \equiv 1 \pmod{p^2}$ for $p = 2$? for $p = 3$? for $p = 5$? for $p = 7$? for $p = 11$? (After completing Exercise 13 you may be interested to learn that $2^{1092} \equiv 1 \pmod{1093^2}$.)

14. For each m greater than 1, how many primes are there in the closed interval $[m!+2, m!+m]$?

15. Suppose p denotes a prime congruent to 3 modulo 4; use Wilson's theorem to prove that

$$\left(\frac{p-1}{2}\right)!^2 \equiv 1 \pmod{p}.$$

[Hint: See the hint for Exercise 9.]

16. Find the *least* positive integer n that satisfies each of the following congruences.

(a) $3^{36} \equiv n \pmod{7}$

(b) $7^{38} \equiv n \pmod{11}$

(c) $7^{128} \equiv n \pmod{13}$.

17. Decide whether or not 17 is a prime by determining whether 16! is congruent to -1 modulo 17. Is this an efficient test for determining whether or not 1093 is a prime?

18. Let $\lambda(m)$ be the least positive integer such that for each integer a relatively prime to m,

$$a^{\lambda(m)} \equiv 1 \pmod{m}.$$

Compute $\lambda(m)$ for each m $(1 \leq m \leq 10)$. *Is* $\lambda(m) \leq \phi(m)$ for each m? Is $\lambda(m) < \phi(m)$ for some m?

19. In 500 B.C. the Chinese seem to have known that $2^p \equiv 2 \pmod{p}$ for each prime p. They also assumed that if $2^n \equiv 2 \pmod{n}$, then n is prime.

(a) Show that 341 is not a prime.

(b) Show that $2^{10} \equiv 1 \pmod{341}$.

(c) Show that $2^{341} \equiv 2 \pmod{341}$.

20. Let $\Phi_n = 2^{2^n} + 1$.

(a) Prove that $\Phi_n \mid (2^{2^{n+1}} - 1)$.

(b) Prove that if $a \mid b$, then $(2^a - 1) \mid (2^b - 1)$.

(c) Prove that $(2^{2^{n+1}} - 1) \mid (2^{2^{2^n}} - 1)$.

(d) Prove that $(2^{2^{2^n}} - 1) \mid (2^{2^{2^n}+1} - 2)$.

(e) Prove that $\Phi_n \mid (2^{\Phi_n} - 2)$.

(f) P. Fermat observed that $\Phi_1 = 5$, $\Phi_2 = 17$, $\Phi_3 = 257$, and $\Phi_4 = 65537$ are all primes, and he suggested that Φ_n is prime for each n. Discuss Fermat's suggestion in the light of Exercises 19 and 20(e).

21. Prove that if p denotes an odd prime, then

$$2^{\frac{p-1}{2}} \equiv \pm 1 \pmod{p}.$$

22. Prove that if $p \equiv 3 \pmod{4}$, then the product of all the odd integers less than p is congruent either to 1 or to -1 modulo p. [Hint: If $p = 2l + 1$, then $(p - 1)! = 2^l l! \cdot 1 \cdot 3 \cdot 5 \cdot \ldots \cdot (2l - 1)$.]

23. Prove that if $p \equiv 3 \pmod{4}$, then the product of all the even integers less than p is congruent either to 1 or to -1 modulo p.

5–3 THE CHINESE REMAINDER THEOREM

Having considered single linear congruences in Section 5–1, we now turn to the problem of finding solutions of systems of linear congruences. A solution of the system of congruences

$$a_1 x \equiv b_1 \pmod{m_1},$$
$$a_2 x \equiv b_2 \pmod{m_2},$$
$$\ldots,$$

and

$$a_s x \equiv b_s \pmod{m_s}$$

is an integer that satisfies each congruence in the system.

The simplest examples of such systems arise in the solution of single linear congruences with large moduli. Let m have the prime factorization

$$m = p_1^{e_1} p_2^{e_2} \ldots p_s^{e_s};$$

then, as a consequence of the fundamental theorem of arithmetic, $m \mid n$ if and only if $p_i^{e_i} \mid n$ for each i. Hence,

$$A \equiv B \pmod{m}$$

if and only if all of the congruences

$$A \equiv B \pmod{p_1^{e_1}},$$
$$A \equiv B \pmod{p_2^{e_2}},$$
$$\ldots,$$

and

$$A \equiv B \pmod{p_s^{e_s}}$$

hold. It follows that the congruence

$$ax \equiv b \pmod{m} \tag{5-3-1}$$

has the same set of solutions as the system of simultaneous congruences

$$ax \equiv b \pmod{p_1^{e_1}},$$
$$ax \equiv b \pmod{p_2^{e_2}},$$
$$\ldots, \tag{5-3-2}$$

and

$$ax \equiv b \pmod{p_s^{e_s}}.$$

Although there are several congruences to solve in (5–3–2), their moduli are generally much smaller than m, and, as we shall see, computations are thus simplified.

Example 5–6: Let us replace the congruence

$$3x \equiv 11 \pmod{2275}$$

by a system of linear congruences with smaller moduli. Since $2275 = 5^2 \cdot 7 \cdot 13$, our congruence may be replaced by the system

$$3x \equiv 11 \pmod{25}, \tag{5-3-3}$$

$$3x \equiv 11 \pmod{7}, \tag{5-3-4}$$

and

$$3x \equiv 11 \pmod{13}, \tag{5-3-5}$$

The basic existence theorem for systems of congruences is the Chinese Remainder Theorem, which has this name because Sun-Tsu is believed to be the first mathematician who studied special cases of this theorem. The proof we shall give furnishes us with a method for constructing solutions of a system of congruences from solutions of the individual congruences in the system.

THEOREM 5–4: *Suppose $m_1, m_2, \ldots, m_s$ are s integers, no two of which have a common factor other than 1. Let $M = m_1 m_2 \ldots m_s$, and suppose that $a_1, a_2, \ldots, a_s$ are integers such that* g.c.d.$(a_i, m_i) = 1$ *for each i. Then the s congruences*

$$a_1 x \equiv b_1 \pmod{m_1},$$

$$a_2 x \equiv b_2 \pmod{m_2},$$

$$\ldots,$$

and

$$a_s x \equiv b_s \pmod{m_s}$$

have a simultaneous solution that is unique modulo M.

PROOF: From the solutions of each particular congruence, we shall construct one common to the entire set. First we choose integers $c_1, c_2, \ldots, c_s$ such that

$$a_i c_i \equiv b_i \pmod{m_i}$$

(such integers c_i exist by Theorem 5–1). Now let $n_i = M/m_i$. Since no two of the m_i have a common factor, we see that g.c.d.$(n_i, m_i) = 1$. Hence, by Corollary 5–1, there exists an $\tilde{n}_i$ such that $n_i \tilde{n}_i \equiv 1 \pmod{m_i}$.

We now show that the number x_0 defined by

$$x_0 = c_1 n_1 \tilde{n}_1 + c_2 n_2 \tilde{n}_2 + \ldots + c_s n_s \tilde{n}_s$$

is a solution of the original system of s congruences. First note that m_i divides each n_j except for n_i. Thus

$$\begin{aligned} a_i x_0 &= a_i c_1 n_1 \tilde{n}_1 + a_i c_2 n_2 \tilde{n}_2 + \ldots + a_i c_s n_s \tilde{n}_s \\ &\equiv a_i c_i n_i \tilde{n}_i \pmod{m_i} \\ &\equiv a_i c_i \pmod{m_i} \\ &\equiv b_i \pmod{m_i}. \end{aligned}$$

Hence, x_0 is a solution of each congruence.

If y is a solution of the system of s congruences, then, by Theorem 5-1, $x_0 \equiv c_i \equiv y \pmod{m_i}$. Hence $m_i \mid x_0 - y$ for each m_i; since no two of the m_i have a common factor, $m_1 m_2 \ldots m_s \mid x_0 - y$, that is, $M \mid x_0 - y$. Thus $y \equiv x_0 \pmod{M}$.

Example 5-7: We now use the Chinese Remainder Theorem to solve the problem posed in the preface to this chapter; namely, what is the least positive x such that

$$x \equiv 2 \pmod{3},$$

$$x \equiv 3 \pmod{5},$$

and

$$x \equiv 2 \pmod{7}?$$

In terms of Theorem 5-4, $a_1 = a_2 = a_3 = 1$; $c_1 = 2$, $c_2 = 3$, and $c_3 = 2$; $m_1 = 3$, $m_2 = 5$, and $m_3 = 7$; $M = 105$; $n_1 = 35$, $n_2 = 21$, and $n_3 = 15$. Now $35\bar{n}_1 \equiv 1 \pmod{3}$, that is, $2\bar{n}_1 \equiv 1 \pmod{3}$; thus we may let $\bar{n}_1 = 2$. Next, $21\bar{n}_2 \equiv 1 \pmod{5}$, and so we may let $\bar{n}_2 = 1$. Finally, $15\bar{n}_3 \equiv 1 \pmod{7}$, and we may let $\bar{n}_3 = 1$. Consequently, a solution of the system is

$$\begin{aligned} x_0 &= 2 \cdot 35 \cdot 2 + 3 \cdot 21 \cdot 1 + 2 \cdot 15 \cdot 1 \\ &= 140 + 63 + 30 = 233. \end{aligned}$$

We conclude that all solutions are congruent to 233 (mod 105), and that the least positive solution is 23.

Example 5-8: Let us now solve the congruence considered in Example 5-6. Since g.c.d.(3,2275) = 1, the congruence has a solution. We note that $x = 12$ is a solution of (5-3-3); $x = 6$, a solution of (5-3-4); and $x = 8$, a solution of (5-3-5). (These particular solutions can easily be obtained by inspection.) Using the notation of Theorem 5-4, we see that

$$c_1 = 12,$$

$$c_2 = 6,$$

and

$$c_3 = 8.$$

Next we must find $\bar{n}_1$, $\bar{n}_2$, and $\bar{n}_3$ such that

$$\frac{2275}{25}\bar{n}_1 = 91\bar{n}_1 \equiv 1 \pmod{25},$$

$$\frac{2275}{7}\tilde{n}_2 = 325\tilde{n}_2 \equiv 1 \pmod 7,$$

and

$$\frac{2275}{13}\tilde{n}_3 = 175\tilde{n}_3 \equiv 1 \pmod{13}.$$

Equivalently, we must solve the simpler congruences

$$16\tilde{n}_1 \equiv 1 \pmod{25},$$

$$3\tilde{n}_2 \equiv 1 \pmod 7,$$

and

$$6\tilde{n}_3 \equiv 1 \pmod{13}.$$

By inspection, we see that $\tilde{n}_1 = 11$, $\tilde{n}_2 = 5$, and $\tilde{n}_3 = 11$ are solutions.

We now use the foregoing results to obtain a solution of the original congruence

$$3x \equiv 11 \pmod{2275}.$$

It is

$$\begin{aligned} x_0 &= 12 \cdot 91 \cdot 11 + 6 \cdot 325 \cdot 5 + 8 \cdot 175 \cdot 11 \\ &= 12012 + 9750 + 15400 \\ &= 37162 \equiv 762 \pmod{2275}. \end{aligned}$$

Hence we see that 762 is the smallest possible solution of the congruence.

In summary, we observe that to arrive at this answer we had to solve six congruences; however, the moduli involved were small, so that each of the six congruences was easy to solve by inspection.

EXERCISES

1. Find *all* solutions of each of the following systems of congruences:

(a) $x \equiv 1 \pmod 2$
$x \equiv 2 \pmod 3$
$x \equiv 3 \pmod 5$

(b) $x \equiv 1 \pmod 3$
$x \equiv 3 \pmod 5$
$x \equiv 5 \pmod 7$

(c) $3x \equiv 1 \pmod 5$
$4x \equiv 6 \pmod{14}$
$5x \equiv 11 \pmod 3$

(d) $4x \equiv 2 \pmod 6$
$3x \equiv 5 \pmod 7$
$2x \equiv 4 \pmod{11}$

(e) $x \equiv 1 \pmod{3}$
$x \equiv 2 \pmod{4}$
$x \equiv 3 \pmod{7}$
$x \equiv 4 \pmod{11}$

(f) $2x \equiv 1 \pmod{5}$
$3x \equiv 9 \pmod{6}$
$4x \equiv 1 \pmod{7}$
$5x \equiv 9 \pmod{11}$.

2. Find the least positive integer that yields the remainders 1, 3, and 5 when divided by 5, 7, and 9, respectively.

3. In Exercise 2 of Section 8–1, it is asserted that if $\Phi_n = 2^{2^n} + 1$, then g.c.d.$(\Phi_n, \Phi_m) = 1$ whenever $n \neq m$. Using this assertion, prove that for each n there exist n consecutive integers $N, N+1, \ldots, N+n-1$ such that the first is divisible by Φ_1; the second, by Φ_2; the third, by Φ_3; . . .; and the nth, by Φ_n. When $n = 2$, what is the least possible value of N? When $n = 3$?

4. Prove that for each n there exists a sequence of n consecutive integers each of which is divisible by a perfect square exceeding 1.

5. Find three consecutive integers the first of which is divisible by the square of a prime; the second, by the cube of a prime; and the third, by the fourth power of a prime.

6. A class in number theory was to divide itself into groups of equal sizes to study the Chinese Remainder Theorem. When the class was divided into groups of 3, two students were left out; when into groups of 4, one was left out. When it was divided into groups of five, the students found that if the professor was added to one of the groups, no one was left out. Since the professor had never really understood the Chinese Remainder Theorem when he was in college, the last arrangement worked out nicely. How many students were there in the class?

5-4 POLYNOMIAL CONGRUENCES

After a study of linear congruences, it is natural to examine congruences of the form

$$f(x) \equiv 0 \pmod{m},$$

where

$$f(x) = a_0x^n + a_1x^{n-1} + \ldots + a_n$$

($a_0 \neq 0$; all the a_i are integers). The function $f(x)$ is called a *polynomial* of *degree n* with *integral coefficients.*

We shall not delve extensively into polynomial congruences; however, the subject contains an important theorem of Lagrange that we shall need when we study primitive roots.

THEOREM 5–5: *If $f(x)$ is a polynomial of degree n with integral coefficients (that is, if $f(x) = a_0x^n + a_1x^{n-1} + \ldots + a_n$), and if p is a prime such that $p \nmid a_0$, then the congruence*

$$f(x) \equiv 0 \pmod p$$

has at most n mutually incongruent solutions modulo p.

PROOF: We proceed by mathematical induction on the degree n. If $n=0$, then $f(x) = a_0$. Since $a_0 \not\equiv 0 \pmod p$, there are no solutions. If $n = 1$, then $f(x) = a_0x + a_1$; and we know by Theorem 5–1 that the congruence

$$a_0x \equiv -a_1 \pmod p$$

has exactly one solution (mod p), because g.c.d.$(a_0, p) = 1$.

Assume now that we have proved the theorem for all polynomials of degree n less than k, where $k \geq 2$. Suppose that $f(x) = a_0x^k + \ldots + a_k$ is a polynomial of degree k with $k+1$ mutually incongruent solutions modulo p, say $w_1, w_2, \ldots, w_{k+1}$. We define a polynomial

$$g(x) = f(x) - a_0(x - w_1)(x - w_2) \ldots (x - w_k).$$

Now $g(x) \equiv f(x) \equiv 0 \pmod p$ if $x = w_1, w_2, \ldots, w_k$. Furthermore, $g(x)$ is a polynomial of degree less than k, because the leading term a_0x^k of $f(x)$ is cancelled by the term a_0x^k in the product on the right. Thus, since the number of solutions of the congruence $g(x) \equiv 0 \pmod p$ is larger than the degree of $g(x)$, it follows from the induction hypothesis that

$$g(x) \equiv 0 \pmod p$$

for each integer x. In particular,

$$g(w_{k+1}) \equiv 0 \pmod p.$$

Hence

$$\begin{aligned} 0 &\equiv g(w_{k+1}) \pmod p \\ &\equiv f(w_{k+1}) - a_0(w_{k+1}-w_1)(w_{k+1}-w_2)\dots(w_{k+1}-w_k) \pmod p \\ &\equiv -a_0(w_{k+1}-w_1)(w_{k+1}-w_2)\dots(w_{k+1}-w_k) \pmod p. \end{aligned}$$

Thus $p \mid a_0(w_{k+1}-w_1)(w_{k+1}-w_2)\ \dots\ (w_{k+1}-w_k)$, so that, by Corollary 2–4, p divides one of the factors. However, $p \nmid a_0$, and since the w_i are mutually incongruent, the difference between any two w_i is not divisible by p. This contradiction leads us to conclude that the congruence $f(x) \equiv 0 \pmod p$ has at most k mutually incongruent solutions, and our theorem is established. ■

EXERCISES

1. Find *all* solutions of each of the following congruences.

   (a) $x^2 + x + 1 \equiv 0 \pmod{11}$

   (b) $x^3 + x + 1 \equiv 0 \pmod{13}$

   (c) $x^4 + x^3 + 2 \equiv 0 \pmod 7$.

2. Prove that if $f(x) = a_0x^n + a_1x^{n-1} + \dots + a_n$ and m is an integer, then $k! \mid f^{(k)}(m)$, where $f^{(k)}$ is the kth derivative of f.

3. From calculus recall that the Taylor series for the polynomial

   $$f(x) = a_0x^n + a_1x^{n-1} + \dots + a_n$$

   is

   $$f(x+h) = f(x) + f'(x)h + \frac{f''(x)h}{2!} + \dots + \frac{f^{(n)}(x)h^n}{n!}.$$

   Prove that if $a_0, a_1, \dots, a_n, m, r$, are integers, and p is a prime, then

   $$f(m + rp^s) \equiv f(m) + rp^sf'(m) \pmod{p^{s+1}}.$$

4. Suppose that $f(m) \equiv 0 \pmod{p^s}$ and that $p \nmid f'(m)$, where p is a prime. Use Exercise 3 to prove that there exists an r (unique modulo p) such that

   $$f(m + rp^s) \equiv 0 \pmod{p^{s+1}}.$$

5. Use Exercise 4 to determine the number of solutions of the congruence

$$x^7 + x + 1 \equiv 0 \pmod{343}.$$

6. A polynomial is said to be *monic* if its leading coefficient is 1. A monic polynomial $f(x)$ of degree n is called *reducible* modulo p if there exist nonconstant, monic polynomials $g(x)$ and $h(x)$, each of degree less than n, such that

$$f(x) \equiv g(x)h(x) \pmod{p}.$$

Is $x^4 + 1$ reducible modulo 2? modulo 3? modulo 5? modulo 7? modulo 11? modulo 13? Contrast your answers with the fact that $x^4 + 1$ is irreducible in the sense of elementary algebra.

7. Use Euler's theorem and Theorem 5–5 to prove that

$$x^{p-1} - 1 \equiv (x-1)(x-2) \ldots (x-p+1) \pmod{p}$$

for each integer x and each prime p.

8. What may one deduce from the result given in Exercise 7 by setting $x = p$?

CHAPTER 6

ARITHMETIC FUNCTIONS

Any function whose domain is some set of integers is called an *arithmetic function.* In Chapters 4 and 5, we encountered $\phi(n)$, the Euler ϕ-function. Other arithmetic functions we shall meet are the number $d(n)$ of divisors of n, and the sum $\sigma(n)$ of the divisors of n.

We shall use combinatorial techniques (see Chapter 3) to establish multiplicative and other properties of $\phi(n)$, $d(n)$, and $\sigma(n)$. Then, we shall generalize our results by means of the Möbius function $\mu(n)$.

6–1 COMBINATORIAL STUDY OF $\phi(n)$

In Chapter 4, we met the Euler ϕ-function $\phi(n)$ in connection with reduced residue systems modulo n. Recall that $\phi(n)$ is defined as the number of positive integers not exceeding n that are relatively prime to n.

Example 6–1: In Example 4–6, we showed that $\phi(p) = p - 1$ for any prime p. It is almost as easy to determine $\phi(p^n)$ for each positive integer n and each prime p. We proceed by counting the positive integers not exceeding p^n that are *not relatively prime* to p^n. Clearly, an integer m has a factor greater than 1 in common with p^n if and only if $p \mid m$. Hence, the positive integers less than or equal to p^n that are not relatively prime to p^n are the p^{n-1} multiples of p: $p, 2p, 3p, \ldots, p^{n-1} \cdot p$. Therefore, there are $p^n - p^{n-1}$ positive integers not exceeding p^n that are relatively prime to p^n. That is,

$$\phi(p^n) = p^n - p^{n-1} = p^n\left(1 - \frac{1}{p}\right).$$

For integers other than prime powers, the values of the ϕ-function are seemingly irregularly distributed (see Table 6–1); nevertheless, in Theorem 6–2 we shall be able to obtain an explicit formula for $\phi(n)$ by generalizing the combinatorial procedure of Example 6–1.

Example 6–2: From Example 6–1, we see that

$$1 + \phi(p) + \phi(p^2) + \ldots + \phi(p^n) = 1 + (p-1) + (p^2 - p) + \ldots + (p^n - p^{n-1}) = p^n,$$

since the terms in the penultimate expression mutually cancel each other save for p^n.

In Theorem 6–1, we shall establish an important formula involving the Euler ϕ-function that generalizes Example 6–2.

THEOREM 6–1: $\sum_{d \mid n} \phi(d) = n.$

NOTATION: $\sum_{d \mid n} \phi(d)$ denotes the sum of the values of the ϕ-function taken for all the divisors of n. For example, when $n = 6$,

$$\sum_{d \mid 6} \phi(d) = \phi(1) + \phi(2) + \phi(3) + \phi(6) = 1 + 1 + 2 + 2 = 6.$$

PROOF: We use a combinatorial argument. Let S_n denote the set $\{1,2,\ldots,n\}$. If we let $\#(S_n)$ denote the number of elements of S_n, then clearly

$$\#(S_n) = n.$$

For each d that divides n, we denote by $T_d(n)$ the set of positive integers not exceeding n whose greatest common divisor with n is d. Taking $n = 6$, for example, we see that $T_1(6) = \{1,5\}$, $T_2(6) = \{2,4\}$, $T_3(6) = \{3\}$, and $T_6(6) = \{6\}$. Clearly, for each n the various $T_d(n)$ have no common elements. Furthermore, for any $m \in S_n$, we see that $m \in T_d(n)$ where $d =$ g.c.d.(m,n). Consequently,

$$n = \#(S_n) = \sum_{d \mid n} \#\{T_d(n)\}.$$

We now show that $T_d(n)$ has $\phi(n/d)$ elements. First note that all elements of $T_d(n)$ are multiples of d, and are less than or equal to n. Thus the elements of $T_d(n)$ are found among the numbers $d, 2d, \ldots, (n/d)d$. Now if g.c.d.$(a,n/d) = e$, then clearly g.c.d.$(ad,n) = ed$, and $ed = d$ if and only if $e = 1$. Therefore, the only numbers ad in $T_d(n)$ are those for which g.c.d.$(a,n/d) = 1$, there being $\phi(n/d)$ such

TABLE 6-1: VALUES OF $\phi(n)$, $d(n)$, AND $\sigma(n)$ $(1 \leq n \leq 20)$.

n	$\phi(n)$	$d(n)$	$\sigma(n)$
1	1	1	1
2	1	2	3
3	2	2	4
4	2	3	7
5	4	2	6
6	2	4	12
7	6	2	8
8	4	4	15
9	6	3	13
10	4	4	18
11	10	2	12
12	4	6	28
13	12	2	14
14	6	4	24
15	8	4	24
16	8	5	31
17	16	2	18
18	6	6	39
19	18	2	20
20	8	6	42

numbers, by definition of the ϕ-function. Hence

$$n = \#(S_n) = \sum_{d \mid n} \#\{T_d(n)\} = \sum_{d \mid n} \phi(n/d).$$

We note that

$$\sum_{d \mid n} \phi(n/d) = \sum_{d \mid n} \phi(d),$$

for as d assumes the values of the various divisors of n, so does the complementary divisor n/d. Thus we have proved our theorem. ■

We now define the Möbius function $\mu(n)$, which will be useful in the next theorem.

DEFINITION 6-1:

$$\mu(n) = \begin{cases} 1 & \text{if } n = 1, \\ 0 & \text{if } p^2 \mid n \text{ for some prime } p, \\ (-1)^r & \text{if } n = p_1 p_2 \ldots p_r \text{ where the } p_i \text{ are distinct primes.} \end{cases}$$

Example 6-3: $\mu(2) = -1$, $\mu(3) = -1$, $\mu(4) = 0$, $\mu(5) = -1$, and $\mu(6) = 1$.

The first part of the following theorem has formal similarities to Theorem 6–1.

Theorem 6–2: $\phi(n) = \sum_{d \mid n} \mu(d) \frac{n}{d} = n \prod_{p \mid n} \left(1 - \frac{1}{p}\right).$

Notation: $\prod_{p \mid n} \left(1 - \frac{1}{p}\right)$ denotes the product of all numbers of the form $\left(1 - \frac{1}{p}\right)$ with p taking as values the distinct prime divisors of n. Sometimes we may use the symbol $\prod_{i=1}^{n} A_i$; this denotes the product $A_1 A_2 \ldots A_n$.

Proof: We proceed by mathematical induction on the number of prime factors of n. If n has one prime factor, say $n = q^\alpha$, then, by Example 6–1,

$$\phi(n) = \phi(q^\alpha) = q^\alpha - q^{\alpha - 1}.$$

On the other hand, substitution of the definition of n gives us two equalities:

$$\begin{aligned} \sum_{d \mid n} \mu(d) \frac{n}{d} &= \mu(1) q^\alpha + \mu(q) q^{\alpha-1} + \mu(q^2) q^{\alpha-2} + \ldots + \mu(q^\alpha) \\ &= q^\alpha - q^{\alpha-1} + 0 + \ldots + 0 \\ &= q^\alpha - q^{\alpha-1}, \end{aligned}$$

and

$$n \prod_{p \mid n} \left(1 - \frac{1}{p}\right) = q^\alpha \left(1 - \frac{1}{q}\right) = q^\alpha - q^{\alpha-1}.$$

Consequently, when n has one prime factor, all three of the expressions in the statement of Theorem 6–2 have the same value.

Now let us assume that the theorem is true for each integer with k or fewer distinct prime factors. Suppose $n = n'p^\alpha$, where n' has k distinct prime factors and p is a prime that does not divide n'. Then, by the induction hypothesis,

$$\phi(n') = \sum_{d \mid n'} \mu(d) \frac{n'}{d} = n' \prod_{p \mid n'} \left(1 - \frac{1}{p}\right).$$

Let us now divide the set $\{1, 2, \ldots, n\}$ into p^α subsets, each consisting of n' consecutive integers. Each subset contains $\phi(n')$ integers that are relatively prime to n'. Now, of the $p^\alpha \phi(n')$ positive integers in $\{1, 2, \ldots, n\}$ that are relatively prime to n', the only ones having a common factor with n are the $p^{\alpha-1}\phi(n')$ integers that are multiples of p. Hence,

$$\phi(n) = p^{\alpha}\phi(n') - p^{\alpha-1}\phi(n'). \tag{6-1-1}$$

For example, let $n = 30 = 2 \cdot 3 \cdot 5$ and $n' = 6 = 2 \cdot 3$; below, we have underlined the $5 \cdot \phi(6)$ numbers in $\{1,2,\ldots,30\}$ that are relatively prime to 6.

$\underline{1}$ 2 3 4 $\underline{5}$ 6 $\underline{7}$ 8 9 10 $\underline{11}$ 12 $\underline{13}$ 14 15 16 $\underline{17}$ 18

$\underline{19}$ 20 21 22 $\underline{23}$ 24 $\underline{25}$ 26 27 28 $\underline{29}$ 30

To obtain the $5^{1-1}\phi(6) = \phi(6)$ numbers in $\{1,2,\ldots,30\}$ that are relatively prime to 6 but not to 5, we take all of the numbers less than or equal to $30/5 = 6$ that are relatively prime to 6, and multiply each by 5. Thus we obtain $5 \cdot 1 = 5$ and $5 \cdot 5 = 25$. The exclusion of these numbers from those underlined above leaves 1, 7, 11, 13, 17, 19, 23, and 29; that is, all numbers in $\{1,2,\ldots,30\}$ that are relatively prime to 30.

Hence, by (6–1–1),

$$\phi(n) = p^{\alpha}\phi(n') - p^{\alpha-1}\phi(n')$$

$$= p^{\alpha} \sum_{d \mid n'} \mu(d) \frac{n'}{d} - p^{\alpha-1} \sum_{d \mid n'} \mu(d) \frac{n'}{d}$$

$$= \sum_{d \mid n'} \mu(d) \frac{n}{d} - \frac{1}{p} \sum_{d \mid n'} \mu(d) \frac{n}{d}$$

$$= \sum_{\substack{d \mid n \\ p \nmid d}} \frac{\mu(d)n}{d} + \sum_{d \mid n'} \mu(pd) \frac{n}{pd}$$ [here we have used the obvious fact that if $p \nmid d$, then $\mu(pd) = -\mu(d)$]

$$= \sum_{\substack{d \mid n \\ p \nmid d}} \frac{\mu(d)n}{d} + \sum_{\substack{pd \mid n \\ p \nmid d}} \mu(pd) \frac{n}{pd}$$

$$= \sum_{\substack{d \mid n \\ p \nmid d}} \frac{\mu(d)n}{d} + \sum_{\substack{pd \mid n \\ p \nmid d}} \mu(pd) \frac{n}{pd}$$

$$+ \sum_{\substack{p^2d \mid n \\ p \nmid d}} \mu(p^2d) \frac{n}{p^2d} + \ldots + \sum_{\substack{p^{\alpha}d \mid n \\ p \nmid d}} \mu(p^{\alpha}d) \frac{n}{p^{\alpha}d}$$

$$= \sum_{d \mid n} \mu(d) \frac{n}{d}.$$

The penultimate expression may give the appearance of arising from nowhere. Its first two terms are identical with the preceding line; the new terms are equal to zero, since $\mu(N) = 0$ if $q^2 \mid N$ for some prime q, and since, in the new terms, p^2 (or a higher power of p) appears in the argument of μ.

Again from (6–1–1), we see that

$$\begin{aligned}\phi(n) &= p^\alpha \phi(n') - p^{\alpha-1}\phi(n') \\ &= p^\alpha \phi(n')\left(1 - \frac{1}{p}\right) \\ &= p^\alpha n' \left(1 - \frac{1}{p}\right) \prod_{p \mid n'} \left(1 - \frac{1}{p}\right) \\ &= n \prod_{p \mid n} \left(1 - \frac{1}{p}\right).\end{aligned}$$

Thus our theorem is established. ■

Let us derive a new formula for $\phi(n)$ from the second expression in the statement of Theorem 6–2. Suppose that

$$n = p_1^{e_1} p_2^{e_2} \ldots p_s^{e_s},$$

where the p_i are distinct primes; then

$$\begin{aligned}\phi(n) &= n \prod_{p \mid n} \left(1 - \frac{1}{p}\right) \\ &= p_1^{e_1} p_2^{e_2} \ldots p_s^{e_s} \prod_{i=1}^{s} \left(1 - \frac{1}{p_i}\right) \\ &= \prod_{i=1}^{s} (p_i^{e_i} - p_i^{e_i - 1}) \\ &= \prod_{i=1}^{s} \phi(p_i^{e_i}) = \phi(p_1^{e_1})\phi(p_2^{e_2}) \ldots \phi(p_s^{e_s}).\end{aligned} \tag{6-1-2}$$

Thus $\phi(n)$ is the product of the values assigned to the ϕ-function for the prime powers in the factorization of n. This provides an easy way to remember the formula for $\phi(n)$.

Example 6–4: $\phi(120) = \phi(2^3 \cdot 3 \cdot 5)$

$$= \phi(2^3)\phi(3)\phi(5)$$

$$= (2^3 - 2^2) \cdot (3-1) \cdot (5-1)$$
$$= 4 \cdot 2 \cdot 4$$
$$= 32.$$

EXERCISES (* means difficult)

1. Prove that if $\phi(m) \mid m-1$, then there exists no prime p such that $p^2 \mid m$.

*2. Prove that if m is not a prime and $\phi(m) \mid m-1$, then m has at least three distinct prime factors.

*3. Prove that if m is not a prime and $\phi(m) \mid m-1$, then m has at least four distinct prime factors.

4. Prove that $\phi(m)$ is even if $m > 2$.

5. Prove that if n has r distinct prime factors, then $\phi(n) \geq n2^{-r}$.

6. Find all integers n such that $\phi(n) = 12$.

7. C. Goldbach conjectured that every even number greater than 2 is a sum of two primes. P. Erdös conjectured that, for any even number $2n$, there exist integers q and r such that

$$\phi(q) + \phi(r) = 2n.$$

Does the conjecture of Goldbach imply that of Erdös?

8. Suppose g.c.d.$(m,n) = 1$. Prove that if x assumes all values in a reduced residue system modulo m, and y assumes all values in a reduced residue system modulo n, then $nx + my$ assumes all values in a reduced residue system modulo mn.

9. Use Exercise 8 to prove that if g.c.d.$(m,n) = 1$, then $\phi(mn) = \phi(m)\phi(n)$.

10. Use Exercise 9 to give a new proof that

$$\phi(n) = n \prod_{p \mid n} \left(1 - \frac{1}{p}\right).$$

11. Characterize all pairs of integers m and n for which $\phi(mn) = \phi(n)$.

12. R. D. Carmichael conjectured that for each integer n there exists an m different from n such that

$$\phi(n) = \phi(m).$$

(a) Prove Carmichael's conjecture for each n congruent to 2 modulo 4.

(b) Prove Carmichael's conjecture for each n less than 20.

13. Find infinitely many integers n for which $10 \mid \phi(n)$. [Hint: $\phi(11) = 10$.]

14. Prove that there are infinitely many integers n for which $\phi(n)$ is a perfect square.

15. Evaluate $\phi(19)$, $\phi(49)$, $\phi(243)$, and $\phi(1024)$.

6–2 FORMULAE FOR $d(n)$ AND $\sigma(n)$

We now turn our attention to the number $d(n)$ of (positive) divisors of n, and to the sum $\sigma(n)$ of these divisors. As is the case with $\phi(n)$, these functions are easily evaluated when n is a prime power.

Example 6–5: If p denotes a prime, then the positive divisors of p are 1 and p. Thus $d(p) = 2$, and $\sigma(p) = p + 1$. More generally, the positive divisors of p^n are $1, p, p^2, \ldots, p^n$. Hence $d(p^n) = n + 1$, and

$$\sigma(p^n) = 1 + p + p^2 + \ldots + p^n$$

$$= \frac{p^{n+1} - 1}{p - 1} \qquad \text{(by Theorem 1–2).}$$

The following theorem establishes general formulae for $d(n)$ and $\sigma(n)$.

Theorem 6–3: *If $n = p_1^{\alpha_1} \ldots p_s^{\alpha_s}$, then*

$$d(n) = (\alpha_1 + 1)(\alpha_2 + 1) \ldots (\alpha_s + 1),$$

and

$$\sigma(n) = \frac{p_1^{\alpha_1+1} - 1}{p_1 - 1} \frac{p_2^{\alpha_2+1} - 1}{p_2 - 1} \ldots \frac{p_s^{\alpha_s+1} - 1}{p_s - 1}.$$

PROOF: As before, we proceed by mathematical induction on the number of prime factors of n. If n has one prime factor, say $n = p^\alpha$, then

$$d(n) = d(p^\alpha) = \alpha + 1,$$

and

$$\sigma(n) = \sigma(p^\alpha) = \frac{p^{\alpha+1} - 1}{p - 1},$$

by Example 6–5.

Now assume that the theorem is true whenever n has k or fewer distinct prime factors. Let $n = n'p^\alpha$ where n' has k distinct prime factors and p, which is a prime, is not a factor of n'. If

$$n' = p_1^{\alpha_1} p_2^{\alpha_2} \ldots p_k^{\alpha_k},$$

then

$$d(n') = (\alpha_1 + 1)(\alpha_2 + 1) \ldots (\alpha_k + 1)$$

and

$$\sigma(n') = \frac{p_1^{\alpha_1+1} - 1}{p_1 - 1} \frac{p_2^{\alpha_2+1} - 1}{p_2 - 1} \frac{p_k^{\alpha_k+1} - 1}{p_k - 1}.$$

Let $d_1, d_2, \ldots, d_s$ denote the $d(n')$ divisors of n'. Then the divisors of n are $d_1, d_2, \ldots, d_s, pd_1, pd_2, \ldots, pd_s, p^2d_1, p^2d_2, \ldots, p^2d_s, \ldots, p^\alpha d_1, p^\alpha d_2, \ldots, p^\alpha d_s$. Thus

$$\begin{aligned} d(n) &= d(n')(\alpha + 1) \\ &= (\alpha_1 + 1)(\alpha_2 + 1) \ldots (\alpha_k + 1)(\alpha + 1), \end{aligned}$$

and similarly,

$$\begin{aligned} \sigma(n) &= \sigma(n') + p\sigma(n') + \ldots + p^\alpha\sigma(n') \\ &= \sigma(n')(1 + p + \ldots + p^\alpha) \\ &= \sigma(n') \frac{p^{\alpha+1} - 1}{p - 1} \\ &= \frac{p_1^{\alpha_1+1} - 1}{p_1 - 1} \frac{p_2^{\alpha_2+1} - 1}{p_2 - 1} \frac{p_k^{\alpha_k+1} - 1}{p_k - 1} \frac{p^{\alpha+1} - 1}{p - 1}. \end{aligned}$$

Thus, our theorem follows by mathematical induction. ■

COROLLARY 6–1: *If* $n = p_1^{\alpha_1} p_2^{\alpha_2} \ldots p_s^{\alpha_s}$, *then*

$$d(n) = d(p_1^{\alpha_1}) d(p_2^{\alpha_2}) \ldots d(p_s^{\alpha_s})$$

and

$$\sigma(n) = \sigma(p_1^{a_1})\sigma(p_2^{a_2}) \ldots \sigma(p_s^{a_s}).$$

Example 6–6: $d(120) = d(2^3 \cdot 3 \cdot 5) = d(2^3)d(3)d(5) = 4 \cdot 2 \cdot 2 = 16$, and $\sigma(120) = \dfrac{2^4 - 1}{2 - 1} \cdot \dfrac{3^2 - 1}{3 - 1} \cdot \dfrac{5^2 - 1}{5 - 1} = 15 \cdot 4 \cdot 6 = 360.$

EXERCISES (* means difficult)

1. Prove that $d(n)$ is odd if and only if n is a perfect square.

2. Prove that $\prod_{d \mid n} d = n^{d(n)/2}$.

3. Prove that if $n = p_1^{a_1}p_2^{a_2} \ldots p_r^{a_r}$, then

$$\sigma(n)\phi(n) = n^2(1 - p_1^{-a_1-1}) \ldots (1 - p_r^{-a_r-1}),$$

and

$$\phi(n)\sigma(n) > n^2\left(1 - \frac{1}{p_1^2}\right)\left(1 - \frac{1}{p_2^2}\right) \ldots \left(1 - \frac{1}{p_r^2}\right).$$

4. Prove that the number of ways of writing n as a sum of consecutive integers equals $d(m)$ where m is the largest odd number dividing n. (For example, if $n = 15$, then 15, $7+8$, $4+5+6$, and $1+2+3+4+5$ are the ways of writing 15 as a sum of consecutive integers, and therefore $d(15) = 4$.)

5. Prove that $\sigma(n) \equiv d(m) \pmod 2$ where m is the largest odd factor of n.

*6. Prove that if $n = p_1^{a_1}p_2^{a_2} \ldots p_r^{a_r}$, then

$$\sum_{d \mid n} d\phi(d) = \frac{p_1^{2a_1+1} + 1}{p_1 + 1} \cdots \frac{p_r^{2a_r+1} + 1}{p_r + 1}$$

*7. Prove that if $n = p_1^{a_1}p_2^{a_2} \ldots p_r^{a_r}$, then

$$\sum_{e \mid n} ed(e)$$

$$= \prod_{j=1}^{r} \{(a_j+1)p_j^{a_j+2} - (a_j+2)p_j^{a_j+1} + 1\}(p_j - 1)^{-2}.$$

8. Prove that $d(n) < 2\sqrt{n}$. [Hint: Prove that each factorization of n has one factor that does not exceed $\sqrt{n}$.]

9. If $\sigma(n) = 2n$, n is a perfect number. Prove that if n is a perfect number, then $\sum_{d \mid n} \frac{1}{d} = 2$.

10. Evaluate $\sigma(210)$, $\phi(100)$, and $\sigma(999)$.

11. Evaluate $d(47)$, $d(63)$, and $d(150)$.

12. Prove that $\frac{\phi(n)\sigma(n)+1}{n}$ is an integer if n is a prime, and that it is not an integer if n is divisible by the square of a prime.

13. Let $S(n)$ denote the number of positive integers not exceeding n that are square-free (that is, not divisible by the square of a prime). Compare $S(n)$ and $\frac{\phi(n)\sigma(n)+1}{n}$ for $n \leq 20$. Can you conjecture which is always the larger?

14. Show that Goldbach's conjecture (see Exercise 7 of Section 6–1) implies that for each even number $2m$ there exist integers m_1 and m_2 such that

$$\sigma(m_1) + \sigma(m_2) = 2m.$$

15. Prove that for each integer n, there exist integers n_1 and n_2 such that

$$d(n_1) + d(n_2) = n.$$

6–3 MULTIPLICATIVE ARITHMETIC FUNCTIONS

The four functions we have considered in this chapter belong to the class of multiplicative arithmetic functions.

DEFINITION 6–2: *An arithmetic function $f(n)$ is said to be* **multiplicative** *if*

$$f(mn) = f(m)f(n)$$

whenever g.c.d.$(m,n) = 1$.

THEOREM 6–4: *$\phi(n)$, $d(n)$, $\sigma(n)$, and $\mu(n)$ are multiplicative arithmetic functions.*

PROOF: Suppose g.c.d.$(m,n) = 1$. If

$$n = p_1^{\alpha_1} \dots p_r^{\alpha_r} \text{ and } m = q_1^{\beta_1} \dots q_s^{\beta_s}$$

are the prime factorizations of n and m, then no p_i can occur among the q_i. Hence, by equation (6–1–2),

$$\begin{aligned}\phi(nm) &= \phi(p_1^{\alpha_1})\phi(p_2^{\alpha_2}) \dots \phi(p_r^{\alpha_r})\phi(q_1^{\beta_1})\phi(q_2^{\beta_2}) \dots \phi(q_s^{\beta_s}) \\ &= \phi(n)\phi(m).\end{aligned}$$

By Corollary 6–1,

$$\begin{aligned}d(nm) &= d(p_1^{\alpha_1})d(p_2^{\alpha_2}) \dots d(p_r^{\alpha_r})d(q_1^{\beta_1})d(q_2^{\beta_2}) \dots d(q_s^{\beta_s}) \\ &= d(n)d(m).\end{aligned}$$

Also by Corollary 6–1,

$$\begin{aligned}\sigma(nm) &= \sigma(p_1^{\alpha_1})\sigma(p_2^{\alpha_2}) \dots \sigma(p_r^{\alpha_r})\sigma(q_1^{\beta_1})\sigma(q_2^{\beta_2}) \dots \sigma(q_s^{\beta_s}) \\ &= \sigma(n)\sigma(m).\end{aligned}$$

Finally we note that $\mu(mn) = 0 = \mu(m)\mu(n)$ if any of the exponents exceeds 1; if all the α_i and β_i equal 1, then $\mu(mn) = (-1)^{r+s} = \mu(n)\mu(m)$. ■

EXERCISE

1. Suppose $f(n)$ and $g(n)$ are multiplicative and that $f(p^r) = g(p^r)$ for each r and each prime p. Prove that $f(n) = g(n)$ for all n.

6–4 THE MÖBIUS INVERSION FORMULA

In Section 6–2, we saw two similar formulae related to $\phi(n)$, namely

$$n = \sum_{d \mid n} \phi(d) \quad \text{and} \quad \phi(n) = \sum_{d \mid n} \mu(d)\frac{n}{d}.$$

These two formulae represent a special case of a general theorem on the Möbius function. However, before we can prove this general theorem, we must prove a special result about the Möbius function.

THEOREM 6–5: $\displaystyle\sum_{d \mid n} \mu(d) = \begin{cases} 1 & \textit{if } n = 1, \\ 0 & \textit{if } n > 1. \end{cases}$

PROOF: The assertion is clearly true if $n = 1$. We proceed by mathematical induction on the number of different prime factors of n when $n > 1$.

First, if $n = p^\alpha$, then

$$\begin{aligned}\sum_{d\mid n} \mu(d) &= \mu(1) + \mu(p) + \mu(p^2) + \ldots + \mu(p^\alpha)\\ &= 1 - 1 + 0 + \ldots + 0\\ &= 0.\end{aligned}$$

Suppose the theorem is true for integers with at most k prime factors. Assuming that $n = n'p^\alpha$, where n' has k distinct prime factors and p is a prime that does not divide n', we have the equation

$$\begin{aligned}\sum_{d\mid n} \mu(d) &= \sum_{d\mid n'} \mu(d) + \sum_{d\mid n'} \mu(pd) + \sum_{d\mid n'} \mu(p^2 d) + \ldots + \sum_{d\mid n'} \mu(p^\alpha d)\\ &= \sum_{d\mid n'} \mu(d) - \sum_{d\mid n'} \mu(d) + 0 + \ldots + 0\\ &= 0.\end{aligned}$$

■

THEOREM 6–6 *(Möbius Inversion Formula): If two arithmetic functions $f(n)$ and $g(n)$ satisfy one of the two conditions*

$$f(n) = \sum_{d\mid n} g(d)$$

and

$$g(n) = \sum_{d\mid n} \mu(d) f\left(\frac{n}{d}\right)$$

for each n, then they satisfy both conditions.

PROOF: First suppose that

$$f(n) = \sum_{d\mid n} g(d);$$

then

$$\begin{aligned}\sum_{d\mid n} \mu(d) f\left(\frac{n}{d}\right) &= \sum_{dd'=n} \mu(d) f(d')\\ &= \sum_{dd'=n} \mu(d) \sum_{e\mid d'} g(e)\end{aligned}$$

$$= \sum_{deh=n} \mu(d)g(e)$$

$$= \sum_{eh'=n} g(e) \sum_{d|h'} \mu(d).$$

By Theorem 6–5, the sum $\sum_{d|h'} \mu(d)$ has the value 0 if $h' > 1$, and the value 1 if $h' = 1$. Hence

$$\sum_{d|n} \mu(d)f\left(\frac{n}{d}\right) = g(n).$$

Conversely, suppose $g(n) = \sum_{d|n} \mu(d)f\left(\frac{n}{d}\right)$. Then

$$\sum_{d|n} g(d) = \sum_{d|n} \sum_{d'|d} \mu(d')f\left(\frac{d}{d'}\right)$$

$$= \sum_{d'ef=n} \mu(d')f(e)$$

$$= \sum_{eh'=n} f(e) \sum_{d'|h'} \mu(d).$$

As before, Theorem 6–5 implies that the sum $\sum_{d'|h'} \mu(d')$ has the value 0 if $h' > 1$, and the value 1 if $h' = 1$. Hence

$$\sum_{d|n} g(d) = f(n). \qquad \blacksquare$$

Definition 6–3: *If $f(n)$ and $g(n)$ are two arithmetic functions satisfying the condition $f(n) = \sum_{d|n} g(d)$, then we say that $\{f(n), g(n)\}$ is a* **Mobius pair.**

We shall now relate Möbius pairs to multiplicative functions.

Theorem 6–7: *If one of the functions in the Möbius pair $\{f(n), g(n)\}$ is multiplicative, so is the other.*

Proof: Suppose $g(n)$ is multiplicative, and assume g.c.d.$(m,n) = 1$. Then

$$f(mn) = \sum_{d|mn} g(d).$$

Since m and n are relatively prime, we may separate each divisor d of mn into unique factors e and h such that e is a divisor of m, and h

is a divisor of n; note also that g.c.d.$(e,h) = 1$. Hence

$$\begin{aligned} f(mn) &= \sum_{e\mid m} \sum_{h\mid n} g(eh) \\ &= \sum_{e\mid m} g(e) \sum_{h\mid n} g(h) \\ &= f(m)f(n), \end{aligned}$$

and $f(n)$ is also a multiplicative function.

Now suppose $f(n)$ is multiplicative. Then, by Theorem 6–6, we know that

$$g(mn) = \sum_{d\mid mn} \mu(d) f\left(\frac{mn}{d}\right).$$

Hence, following an argument similar to that in the preceding paragraph, we see that

$$\begin{aligned} g(mn) &= \sum_{e\mid m} \sum_{h\mid n} \mu(eh) f\left(\frac{mn}{eh}\right) \\ &= \sum_{e\mid m} \sum_{h\mid n} \mu(e)\mu(h) f\left(\frac{m}{e}\right) f\left(\frac{n}{h}\right) \\ &= \sum_{e\mid m} \mu(e) f\left(\frac{m}{e}\right) \sum_{h\mid n} \mu(h) f\left(\frac{n}{h}\right) \\ &= g(m)g(n), \end{aligned}$$

hence $g(n)$ is multiplicative. ■

We now study the functions $\phi(n)$, $d(n)$, and $\sigma(n)$ by means of Möbius pairs.

THEOREM 6–8: *$\{n,\phi(n)\}$, $\{d(n),1\}$, and $\{\sigma(n),n\}$ are all Möbius pairs.*

PROOF: Theorem 6–1 asserts that

$$n = \sum_{d\mid n} \phi(d),$$

and by definition

$$d(n) = \sum_{d\mid n} 1 \quad \text{and} \quad \sigma(n) = \sum_{d\mid n} d.$$

Thus Theorem 6–8 now follows from Definition 6–3. ■

Noting that the functions $f_1(n) = 1$ and $f_2(n) = n$ are multiplicative, we deduce from Theorems 6–7 and 6–8 that $\phi(n)$, $d(n)$, and $\sigma(n)$ are also multiplicative (an alternative proof of Theorem 6–4).

EXERCISES (* means difficult)

1. Prove that if $\sigma_k(n) = \sum_{d \mid n} d^k$, then $\sigma_k(n)$ is multiplicative.

2. Prove that if $f(n)$ is multiplicative, then

$$\sum_{d \mid n} \mu(d) f(d) = \prod_{p \mid n} \{1 - f(p)\}.$$

3. Use Exercise 2 to give a new proof of the second part of Theorem 6–2; namely, that

$$\sum_{d \mid n} \mu(d) \frac{n}{d} = n \prod_{p \mid n} \left(1 - \frac{1}{p}\right).$$

4. Let $S(n)$ denote the number of square-free integers (see Exercise 13 of Section 6–2) not exceeding n. Prove that

$$S(n) = \sum_{j=1}^{n} |\mu(j)|.$$

*5. With the definition of $S(n)$ given in Exercise 4, prove that

$$S(n) = \sum_{j=1}^{n} \sum_{d^2 \mid j} \mu(d).$$ [Hint: Use Theorem 6–5.]

*6. Use Exercise 5 to prove that

$$S(n) = \sum_{d=1}^{n} \mu(d) \left[\frac{n}{d^2}\right],$$

where $\left[\frac{n}{d^2}\right]$ denotes the largest integer that does not exceed $\frac{n}{d^2}$.

7. Prove that $\dfrac{n}{\phi(n)} = \sum_{d \mid n} \mu^2(d)/\phi(d)$.

8. Prove that $\sum_{d \mid n} \mu(d)\phi(d) = \prod_{p \mid n} (2-p)$.

9. Let $A(d,e)$ be a function defined for all integers d and e. Prove that

$$\sum_{d \mid n} \sum_{e \mid \frac{n}{d}} A(d,e) = \sum_{e \mid n} \sum_{d \mid \frac{n}{e}} A(d,e).$$

10. Prove that if $n = p_1^{a_1} p_2^{a_2} \ldots p_r^{a_r}$, then

$$\sum_{d \mid n} \frac{\phi(d)}{d} = \prod_{i=1}^{r} \left\{1 + a_i\left(1 - \frac{1}{p_i}\right)\right\}.$$

11. Prove that if

$$f(n) = \prod_{d \mid n} g(d),$$

then

$$g(n) = \prod_{d \mid n} f(d)^{\mu\left(\frac{n}{d}\right)}.$$

[Hint: Use logarithms.]

12. Use Exercise 2 of Section 6-2 and Exercise 11 above to prove that

$$n = \prod_{h \mid n} h^{d(h)\mu\left(\frac{n}{h}\right)/2}.$$

13. Prove that if

$$f(x) = \sum_{n=1}^{[x]} g\left(\frac{x}{n}\right),$$

then

$$g(x) = \sum_{n=1}^{[x]} \mu(n) f\left(\frac{x}{n}\right).$$

(The symbol $[x]$ denotes the largest integer not exceeding x.)

14. The Inequality of the Geometric and Arithmetic Means asserts that for any nonnegative real numbers $a_1, a_2, \ldots, a_n$,

$$\frac{a_1 + a_2 + \ldots + a_n}{n} \geq (a_1 a_2 \ldots a_n)^{1/n}.$$

Prove that $\dfrac{\sigma(n)}{d(n)} \geq \prod_{d \mid n} d^{1/d(n)}$.

15. Use Exercise 2 of Section 6–2 and Exercise 14 above to prove that

$$\frac{\sigma(n)}{d(n)} \geq n^{1/2}.$$

CHAPTER 7

PRIMITIVE ROOTS

In Chapter 4, we remarked that the elements of a reduced residue system form a group under the operation of multiplication. Since later chapters require familiarity with the structure of this group, we shall study in some detail the reduced residue system modulo p, where p is a prime. As we proceed, we shall discover an integer g such that $g, g^2, \ldots, g^{p-1}$ constitute a reduced residue system modulo p; if you are familiar with abstract algebra, you can see that such a system is a cyclic group. The integer g is called a *primitive root* modulo p.

7–1 PROPERTIES OF REDUCED RESIDUE SYSTEMS

Example 7–1: We know that $\phi(10) = 4$, and we observe that $\{1,3,7,9\}$ is a reduced residue system modulo 10. Since

$$\begin{aligned} 3^1 &= 3 \equiv 3 \pmod{10}, & 7^1 &= 7 \equiv 7 \pmod{10},\\ 3^2 &= 9 \equiv 9 \pmod{10}, & 7^2 &= 49 \equiv 9 \pmod{10},\\ 3^3 &= 27 \equiv 7 \pmod{10}, & 7^3 &= 343 \equiv 3 \pmod{10},\\ 3^4 &= 81 \equiv 1 \pmod{10}, & 7^4 &= 2401 \equiv 1 \pmod{10}, \end{aligned}$$

we see that each of $\{3,3^2,3^3,3^4\}$ and $\{7,7^2,7^3,7^4\}$ is a reduced residue system modulo 10.

In Example 7–1, we see that 4 $(=\phi(10))$ is the smallest positive integer h for which $g^h \equiv 1 \pmod{10}$, when $g = 3$ or 7.

DEFINITION 7–1: *If h is the smallest positive integer such that*

$$a^h \equiv 1 \pmod{m},$$

we say that a **belongs to the exponent** h modulo m.

THEOREM 7–1: *In order that*

$$a^b \equiv 1 \pmod{m}$$

for some integer b, it is necessary and sufficient that g.c.d.$(a,m) = 1$.

PROOF: Let $d =$ g.c.d.(a,m). Clearly, if $d \mid a$ and $d \mid m$, then $d \mid 1$ and therefore $d = 1$.

If, on the other hand, g.c.d.$(a,m) = 1$, Euler's theorem (Theorem 5–2) asserts that

$$a^{\phi(m)} \equiv 1 \pmod{m}. \qquad \blacksquare$$

THEOREM 7–2: *If a belongs to the exponent h* modulo *m, and*

$$a^r \equiv 1 \pmod{m},$$

then $h \mid r$.

PROOF: By Euclid's division lemma (Theorem 2–1),

$$r = kh + s \ (0 \leq s < h).$$

Hence $1 \equiv a^r \equiv a^{kh+s} \equiv (a^h)^k a^s \equiv a^s \pmod{m}$. Thus $s = 0$, since h is the least positive exponent such that $a^h \equiv 1 \pmod{m}$. Therefore $h \mid r$.

DEFINITION 7–2: *If g is an integer belonging to the exponent* $\phi(m)$ modulo *m, then g is called a* **primitive root** modulo *m.*

THEOREM 7–3: *If g is a primitive root* modulo *m, then g,* $g^2, \ldots, g^{\phi(m)}$ *are mutually incongruent and form a reduced residue system* modulo *m.*

PROOF: Suppose $1 \leq s < r \leq \phi(m)$ and

$$g^r \equiv g^s \pmod{m}.$$

Then $m \mid g^r - g^s$, that is, $m \mid g^s(g^{r-s} - 1)$. Hence, by Theorem 2–3, $m \mid g^{r-s} - 1$. Consequently, $g^{r-s} \equiv 1 \pmod{m}$. Thus, $r - s$ is a positive integer less than $\phi(m)$ such that

$$g^{r-s} \equiv 1 \pmod{m}.$$

This contradicts the fact that g belongs to the exponent $\phi(m)$, and the theorem is proven. $\blacksquare$

Example 7-2: Theorems 7-1 and 7-3 enable us to show that there are no primitive roots modulo 8. Suppose that g is a primitive root; then, since $\phi(8) = 4$, g must belong to the exponent 4 modulo 8. By Theorem 7-1, we know that g.c.d.$(g,8) = 1$; hence g must be congruent to one of 1, 3, 5, and 7 modulo 8. However,

$$1^2 \equiv 1 \pmod 8), \qquad 5^2 \equiv 1 \pmod 8),$$
$$3^2 \equiv 1 \pmod 8), \qquad 7^2 \equiv 1 \pmod 8).$$

Therefore, g cannot belong to the exponent 4 modulo 8, and consequently there are *no* primitive roots modulo 8.

We need not go through trial-and-error calculations, as in Examples 7-1 and 7-2, in order to determine the number of mutually incongruent primitive roots modulo m. There is a simple formula for this number, and this formula is the subject of the remainder of this section.

THEOREM 7-4: *If a belongs to the exponent h* modulo *m and* g.c.d.$(k,h) = d$, *then a^k belongs to the exponent h/d* modulo m.

PROOF: Suppose a^k belongs to the exponent j modulo m. Then

$$a^{kj} \equiv 1 \pmod{m}.$$

Thus $h \mid kj$. Let $h_1 = h/d$ and $k_1 = k/d$; then

$$h_1 \mid k_1 j.$$

Since g.c.d.$(h,k) = d$, it is true that g.c.d.$(h_1,k_1) = 1$; from this we see that

$$h_1 \mid j.$$

On the other hand,

$$a^{kh_1} = a^{h_1 k_1 d} = a^{hk_1} = (a^h)^{k_1} \equiv 1 \pmod{m}.$$

Thus $j \mid h_1$. It follows that $j = h_1 = h/d$. ■

COROLLARY 7-1: *If g is a primitive root* modulo *m*, *then g^r is a primitive root* modulo *m if and only if* g.c.d.$\{r,\phi(m)\} = 1$.

PROOF: By definition, a primitive root is an integer that belongs to the exponent $\phi(m)$. By hypothesis, g belongs to the exponent

$\phi(m)$. Therefore, g^r belongs to $\phi(m)/\text{g.c.d.}\{r,\phi(m)\}$, and this number equals $\phi(m)$ if and only if $\text{g.c.d.}\{r,\phi(m)\} = 1$. ■

THEOREM 7–5: *If there exist any primitive roots* modulo m, *there are exactly* $\phi\{\phi(m)\}$ *mutually incongruent primitive roots.*

PROOF: Let g be a primitive root modulo m. Then $g, g^2, \ldots, g^{\phi(m)}$ form a reduced residue system modulo m. By Corollary 7–1, we know that g^r is a primitive root if and only if $\text{g.c.d.}\{r, \phi(m)\} = 1$. However, by definition of the ϕ-function, exactly $\phi\{\phi(m)\}$ integers in the interval $[1, \phi(m)]$ are relatively prime to $\phi(m)$. ■

Example 7–3: Let $m = 10$. Theorem 7–5 tells us that there are $\phi(\phi(10)) = \phi(4) = 2$ mutually incongruent primitive roots modulo 10 if there are any. From our calculations in Example 7–1, we see that 3 and 7 are two such mutually incongruent primitive roots modulo 10.

Even though we have Theorem 7–5, we are still faced with the problem of determining which moduli have *any* primitive roots at all. This is the subject of the next section.

EXERCISES

1. Let g be a primitive root of m. An *index* of a number a to the base g (written $\text{ind}_g a$) is a number t such that $g^t \equiv a \pmod{m}$. Given that $a \equiv b \pmod{m}$ and that g is a primitive root modulo m, prove the following assertions:

   (a) $\text{ind}_g a \equiv \text{ind}_g b \{\text{mod } \phi(m)\}$,

   (b) $\text{ind}_g ac \equiv \text{ind}_g a + \text{ind}_g c \{\text{mod } \phi(m)\}$,

   (c) $\text{ind}_g a^n \equiv n \, \text{ind}_g a \{\text{mod } \phi(m)\}$.

2. Construct a table of indices of all integers for $m = 17$ and $g = 3$.

3. Solve the congruence $9x \equiv 11 \pmod{17}$, using the table constructed in Exercise 2.

4. Solve the congruence $4y^2 \equiv 1 \pmod{17}$, using the table constructed in Exercise 2.

5. Solve the congruence $12x^2 \equiv 7 \pmod{17}$, using the table constructed in Exercise 2.

6. Find all primitive roots modulo 5, modulo 9, modulo 11, modulo 13, and modulo 15.

7. Suppose g is a primitive root modulo p (a prime) and suppose $m \mid p-1$ $(1 < m < p-1)$. How many integral solutions are there of the congruence

$$x^m - g \equiv 0 \pmod{p}?$$

7-2 PRIMITIVE ROOTS MODULO p

A general theorem asserts that m has primitive roots if and only if m is 2 or 4 or a number of the form p^α or $2p^\alpha$, where p denotes an odd prime. We shall prove only that all primes have primitive roots. The complete theorem is covered in the exercises at the end of this chapter.

THEOREM 7-6: *For each prime p, there exist primitive roots* modulo p.

PROOF: Consider the reduced residue system $1, 2, \ldots, p-1$ modulo p. Let $N(h)$ denote the number of these integers that belong to h modulo p. Now we know that if a belongs to the exponent h modulo p, then $h \mid p-1$. We also know that every element of a reduced residue system belongs to some h modulo p. Consequently,

$$p-1 = \sum_{h \mid p-1} N(h). \tag{7-2-1}$$

We shall now show that $N(h)$ is either 0 or $\phi(h)$. If no integer belongs to h modulo p, then clearly $N(h) = 0$. If a belongs to h modulo p, we examine the equation

$$x^h \equiv 1 \pmod{p}. \tag{7-2-2}$$

By Theorem 5-5, we know that (7-2-2) has at most h mutually incongruent solutions. However, since $a, a^2, \ldots, a^h$ are mutually incongruent solutions of (7-2-2), there are *exactly* h solutions. Thus, any solution of (7-2-2) must be congruent to a^r for some r. From Corollary 7-1, we know that a^r belongs to h if and only if g.c.d.$(r,h) = 1$. Hence, among the $a, a^2, \ldots, a^h$, there are exactly $\phi(h)$ numbers that belong to h modulo p. Thus, in this case, $N(h) = \phi(h)$.

We see, therefore, that $\phi(h) \geq N(h)$ for all h. If there existed a single h for which $\phi(h) > N(h)$, it would follow from (7-2-1) and

Theorem 6–1 that

$$p-1=\sum_{h\mid p-1} N(h) < \sum_{h\mid p-1} \phi(h) = p-1,$$

which is impossible. Hence, $N(h) = \phi(h)$ for all h that divide $p-1$. Thus, $\phi(p-1)$ integers belong to the exponent $p-1$ modulo p; that is, there are $\phi(p-1)$ primitive roots modulo p. ■

Example 7–4: Let $p=5$. A reduced residue system modulo 5 is $\{1,2,3,4\}$. By Theorem 7–6, we know that there are primitive roots modulo 5; hence, by Theorem 7–5, there must be two mutually incongruent primitive roots modulo 5. Since

$$\begin{aligned} 2^1 &\equiv 2 \pmod 5, & 3^1 &\equiv 3 \pmod 5,\\ 2^2 &\equiv 4 \pmod 5, & 3^2 &\equiv 4 \pmod 5,\\ 2^3 &\equiv 3 \pmod 5, & 3^3 &\equiv 2 \pmod 5,\\ 2^4 &\equiv 1 \pmod 5, & 3^4 &\equiv 1 \pmod 5, \end{aligned}$$

we see that 2 and 3 are mutually incongruent primitive roots modulo 5.

EXERCISES (* means difficult)

*1. Suppose g.c.d.$(m,n)=1$, $a^h \equiv 1 \pmod m$, and $a^k \equiv 1 \pmod n$. Prove that if $j = hk/\text{g.c.d.}(h,k)$, then $a^j \equiv 1 \pmod{mn}$.

*2. Prove that if a is odd and $n \geq 3$, then $a^{2^{n-2}} \equiv 1 \pmod{2^n}$. [Hint: Use mathematical induction on n, together with the equality $(a^{2^{n-3}})^2 = a^{2^{n-2}}$.]

*3. Use Exercise 2 to prove that the only powers of 2 having primitive roots are 2 and 4.

4. Prove that g.c.d.$\{\phi(m),\phi(n)\} > 1$ unless either m or n equals 2 or 1. [Hint: Use Exercise 4 of Section 6–1.]

*5. Use Exercises 1 and 4 to prove that an integer divisible by two distinct odd primes cannot have a primitive root.

*6. Use the results in the previous problems to show that the only integers that can have primitive roots are 2, 4, p^r, and $2p^r$, where p is an odd prime.

*7. Prove that if g is a primitive root modulo p (p an odd prime) and $g^{p-1} \equiv 1 \pmod{p^2}$, then $(g+p)^{p-1} \not\equiv 1 \pmod{p^2}$.

8. Use Exercise 7 to prove that there exists a primitive root g modulo p (p an odd prime) such that $g^{p-1} \not\equiv 1 \pmod{p^2}$.

9. Prove that if g.c.d.$(a,p)=1$, then $a^{p^m - p^{m-1}} \equiv 1 \pmod{p^m}$, where p denotes an odd prime.

*10. Let g be a primitive root modulo p (p an odd prime) with $g^{p-1} \not\equiv 1 \pmod{p^2}$. Prove that $g^{(p-1)p^{m-2}} \not\equiv 1 \pmod{p^m}$ for every $m \geq 2$. [Hint: Proceed by mathematical induction on m, using Exercise 9 and the relation $g^{(p-1)p^{m-2}} = (g^{(p-1)p^{m-3}})^p$.]

11. Prove that if g is a primitive root modulo p (p an odd prime), then g belongs to h modulo p^m, where $h = (p-1)p^r$ for some r.

12. Use Exercises 10 and 11 to prove that if g is a primitive root modulo p (p an odd prime) and $g^{p-1} \not\equiv 1 \pmod{p^2}$, then g is a primitive root modulo p^m.

13. Prove that some odd numbers are primitive roots modulo p^m for each odd prime p and each positive integer m.

14. Prove that each odd primitive root modulo p^m (p an odd prime) is a primitive root modulo $2p^m$.

15. How many primitive roots exist for the moduli 6, 7, 8, 9, and 10?

16. To what exponents do 1, 2, 3, 4, 5, 6, 7, 8, 9, and 10 belong modulo 11? In this particular case, verify that $N(h) = \phi(h)$ for all h that divide 10.

CHAPTER 8

PRIME NUMBERS

As you learned in Chapter 2, primes play a fundamental role in the theory of numbers. In fact, some of the most striking results in number theory, the Quadratic Reciprocity Law, Wilson's theorem, and Fermat's little theorem, reveal interesting properties of primes.

Gauss was the first to give careful attention to $\pi(x)$, the number of primes that do not exceed x. Observing that the values of $\pi(x)$ were well approximated by $x/\log x$*, he conjectured that

$$\lim_{x\to\infty} \frac{\pi(x)}{\dfrac{x}{\log x}} = 1. \tag{8-0-1}$$

J. Hadamard and C. de la Vallée Poussin independently proved (8–0–1) in 1896, using some very sophisticated techniques in the theory of complex variables. This result has become known as the Prime Number Theorem. Although, in 1948, Atle Selberg and Paul Erdös proved the Prime Number Theorem without complex variables, theirs and all other proofs known today are intricate and long.

Consequently, after examining some simple results concerning $\pi(x)$, we shall prove a theorem which, while resembling the Prime Number Theorem, is far easier. This result of Tchebychev asserts that there exist positive numbers c_1 and c_2 such that

$$c_1 \frac{x}{\log x} < \pi(x) < c_2 \frac{x}{\log x}, \tag{8-0-2}$$

for all $x \geq 2$.

8–1 ELEMENTARY PROPERTIES OF $\pi(x)$

We have not yet established the existence of infinitely many primes, a problem that is easily settled with Euclid's elegant proof.

*Throughout the text, $\log x = \log_e x$.

THEOREM 8–1: *$\lim_{x\to\infty} \pi(x) = +\infty$; that is, there exist infinitely many primes.*

PROOF: Assume that there are only finitely many primes, say $p_1, p_2, \ldots, p_n$. Let $M = p_1 p_2 \ldots p_n + 1$. Clearly, if we divide M by any of $p_1, p_2, \ldots, p_n$, the remainder is 1. Thus, by our hypothesis that there are only finitely many primes, we deduce that M has no prime factorization. Since this contradicts the fundamental theorem of arithmetic, the only possible conclusion is that there are infinitely many primes. Thus $\lim_{x\to\infty} \pi(x) = +\infty$. ■

On the other hand, it is clear that $\pi(x) \leq x$. Actually $\pi(x)$ is much smaller than x. This is reflected in the following three theorems.

THEOREM 8–2: *If k is any positive integer,*

$$\frac{\pi(x)}{x} \leq \frac{\phi(k)}{k} + \frac{2k}{x}.$$

PROOF: Suppose $[x] = kl + r$, where $0 \leq r < k$, and where $[x]$ denotes the largest integer not exceeding x. We divide the integers of $[1,x]$ into l sets of k consecutive integers plus a remaining set of the r integers $kl+1, kl+2, \ldots, kl+r$.

Among the integers $1, 2, \ldots, k$ there are obviously at most k primes. Among the integers $k+1, k+2, \ldots, 2k$, there are at most $\phi(k)$ primes, since any integer not relatively prime to k has a prime factor in common with k that is less than or equal to k. Similarly, in each of the remaining sets of k consecutive integers, there are at most $\phi(k)$ primes. Finally, in the remaining set of r integers, there are at most r primes. Consequently,

$$\pi(x) \leq k + (l-1)\phi(k) + r \leq 2k + \frac{x}{k}\phi(k).$$

Hence

$$\frac{\pi(x)}{x} \leq \frac{\phi(k)}{k} + \frac{2k}{x}.$$ ■

The next result allows us to estimate the size of $\phi(k)/k$.

THEOREM 8–3: *If $M > 1$ and $p_1, p_2, \ldots, p_s$ are all the primes in $\{1,2,\ldots,M\}$, then*

$$\sum_{n=1}^{M} \frac{1}{n} < \frac{1}{\left(1-\frac{1}{p_1}\right)\left(1-\frac{1}{p_2}\right)\cdots\left(1-\frac{1}{p_s}\right)}.$$

REMARK: From Theorem 8–3, we may deduce as follows that the infinite series $\sum_{i=1}^{\infty} \frac{1}{p_i} = \frac{1}{2} + \frac{1}{3} + \frac{1}{5} + \frac{1}{7} + \frac{1}{11} + \ldots$ diverges: If we let $M \to \infty$ in Theorem 8–3, we see that $\prod_{i=1}^{\infty} \left(1 - \frac{1}{p_i}\right) = 0$. Therefore, $\sum_{i=1}^{\infty} \frac{1}{p_i}$ diverges, by Theorem B–2 of Appendix B.

PROOF: From the formula for the sum of a geometric series, we know that for each prime p

$$\frac{1}{1 - \frac{1}{p}} = 1 + \frac{1}{p} + \frac{1}{p^2} + \frac{1}{p^3} + \ldots.$$

Consequently,

$$\begin{aligned}\frac{1}{\left(1 - \frac{1}{p_1}\right)\left(1 - \frac{1}{p_2}\right) \ldots \left(1 - \frac{1}{p_s}\right)} &= \left(1 + \frac{1}{p_1} + \frac{1}{p_1^2} + \frac{1}{p_1^3} + \ldots\right) \\ &\cdot \left(1 + \frac{1}{p_2} + \frac{1}{p_2^2} + \frac{1}{p_2^3} + \ldots\right) \ldots \left(1 + \frac{1}{p_s} + \frac{1}{p_s^2} + \frac{1}{p_s^3} + \ldots\right) \\ &= \sum_{n \in \Lambda} \frac{1}{n} > \sum_{n \leq M} \frac{1}{n},\end{aligned}$$

where Λ is the set of all integers each of whose prime factors is at most M. That the last equality holds is clear, since multiplying these series together gives all possible terms of the form $\frac{1}{p_1^{\alpha_1} p_2^{\alpha_2} \ldots p_s^{\alpha_s}}$, with $p_i \leq M$. ■

While the next theorem shows that $\pi(x)$ is much smaller than x, we require still more groundwork before proving Tchebychev's result, which explicitly describes the relative smallness of $\pi(x)$.

THEOREM 8–4: $\lim_{x \to \infty} \frac{\pi(x)}{x} = 0.$

PROOF: Since $\frac{\pi(x)}{x} \geq 0$ for $x > 0$, we need only show that we can make $\pi(x)/x$ arbitrarily small by choosing x sufficiently large. Theorem 8–2 has established that

$$\frac{\pi(x)}{x} \leq \frac{\phi(k)}{k} + \frac{2k}{x}$$

for each positive integer k. Let M be a large integer, and let $k = p_1 p_2 \ldots p_s$, where $\{p_1, p_2, \ldots p_s\}$ is the set of all primes not exceeding M. Then

$$\frac{\phi(k)}{k} = \frac{k\left(1 - \frac{1}{p_1}\right)\left(1 - \frac{1}{p_2}\right) \ldots \left(1 - \frac{1}{p_s}\right)}{k}$$

$$= \left(1 - \frac{1}{p_1}\right)\left(1 - \frac{1}{p_2}\right) \ldots \left(1 - \frac{1}{p_s}\right) < \left(\sum_{n=1}^{M} \frac{1}{n}\right)^{-1}.$$

Thus

$$\frac{\pi(x)}{x} \leq \left(\sum_{n=1}^{M} \frac{1}{n}\right)^{-1} + \frac{2p_1 p_2 \ldots p_s}{x}. \tag{8-1-1}$$

It is now a simple matter to make $\pi(x)/x$ as small as we like. Since $\sum_{n=1}^{\infty} \frac{1}{n}$ is a divergent series, we can choose M so large that $\sum_{n=1}^{M} \frac{1}{n} > \frac{2}{\epsilon}$, where $\epsilon > 0$ is an arbitrary positive number. Then, for

$$x > \frac{4p_1 p_2 \ldots p_s}{\epsilon},$$

$$\frac{\pi(x)}{x} < \frac{\epsilon}{2} + \frac{2p_1 p_2 \ldots p_s \epsilon}{4p_1 p_2 \ldots p_s} = \epsilon. \qquad \blacksquare$$

To conclude our section on elementary properties of $\pi(x)$, we now present two additional results indispensable to proving (8–0–2).

THEOREM 8–5: *If* $[x]$ *denotes the largest integer that does not exceed* x, *then* $0 \leq [2x] - 2[x] \leq 1$.

PROOF: The two inequalities

$$2x - 1 < [2x] \leq 2x,$$

and

$$2x - 2 < 2[x] \leq 2x$$

are direct consequences of the definition of $[x]$. Thus

$$-1 < [2x] - 2[x] < 2.$$

However, $[2x]-2[x]$ is an integer, and the only integers in the interval $(-1,2)$ are 0 and 1. Thus $0 \leq [2x]-2[x] \leq 1$. ■

THEOREM 8–6: *If p is a prime, then* $\sum_{j=1}^{\infty}\left[\frac{n}{p^j}\right]$ *is the exponent of p appearing in the prime factorization of* $n!$.

PROOF: Note that if $p > n$, then p does not appear in the prime factorization of $n!$ and every term in $\sum_{j=1}^{\infty}\left[\frac{n}{p^j}\right]$ is zero as desired.

If $p \leq n$, then $\left[\frac{n}{p}\right]$ integers in $\{1,2,\ldots,n\}$ are divisible by p, namely

$$p, 2p, 3p, \ldots, \left[\frac{n}{p}\right]p.$$

Of these integers, $\left[\frac{n}{p^2}\right]$ are again divisible by p:

$$p^2, 2p^2, \ldots, \left[\frac{n}{p^2}\right]p^2.$$

By the same logic, $\left[\frac{n}{p^3}\right]$ of these are divisible by p a third time:

$$p^3, 2p^3, \ldots, \left[\frac{n}{p^3}\right]p^3.$$

After finitely many repetitions of this argument, we see that the total number of times p divides numbers in $\{1,2,\ldots,n\}$ is precisely $\sum_{j=1}^{\infty}\left[\frac{n}{p^j}\right]$; consequently, this sum is the exponent of p appearing in the prime factorization of $n!$. ■

EXERCISES

1. The Fermat numbers are numbers of the form $2^{2^n}+1=\Phi_n$. Prove that if $n < m$, then $\Phi_n \mid \Phi_m - 2$.

2. Prove that if $n \neq m$, then $\text{g.c.d.}(\Phi_n, \Phi_m) = 1$.

3. Use Exercise 2 to give a new proof that there exist infinitely many primes.

4. Modify the proof of Theorem 8–1 to prove that there exist infinitely many primes congruent to 5 (mod 6).

5. Modify the proof of Theorem 8–1 to prove that there exist infinitely many primes congruent to 3 (mod 4).

6. Suppose that $p_1, p_2, \ldots, p_r$ are the only primes congruent to 1 (mod 4). Prove that $4p_1^2p_2^2 \ldots p_r^2 + 1$ is divisible only by primes congruent to 3 (mod 4).

7. Assuming that all odd prime factors of integers of the form $x^2 + 1$ are congruent to 1 (mod 4), use Exercise 6 to prove that there exist infinitely many primes congruent to 1 (mod 4).

8. Let $a_1 = 2$, and for each $n > 1$, let a_n be the least integer for which g.c.d.$(a_i, a_n) = 1$ for each positive $i < n$. What is a_2? a_3? a_4? a_5? a_6? a_n?

9. Use the proof of Theorem 8–1 together with mathematical induction to prove that the nth prime p_n is less than $2^{2^n} + 1$. (Alternatively, use Exercise 3.)

10. Prove (using Exercise 9) that

$$\pi(x) > c \operatorname{loglog} x$$

for some absolute constant c.

11. (a) Prove that every integer can be factored uniquely into the product of a square-free number (see Exercise 13 of Section 6–2) and a perfect square.

    (b) Prove that if only r primes existed, then there would be exactly 2^r square-free numbers.

    (c) From (a) and (b) deduce that if there were only r primes, then there would be at most $2^r\sqrt{n}$ integers not exceeding n. Show that this is impossible for n sufficiently large.

    (d) Deduce from (c) that

$$\pi(x) > c \log x,$$

for some absolute constant c.

12. Let $\theta(x)$ be the sum of the logarithms of all the primes not exceeding x. Prove that

$$\theta(x) \leq \pi(x) \log x.$$

13. Suppose there were only finitely many primes $2,3,5,\ldots,p$. Let M denote the product of all the primes. By evaluating $\phi(M)$, prove that there exist primes not dividing M.

14. Find the smallest positive integer for which x^2-x+41 is not a prime.

15. Find the prime factorization of

$$2{,}432{,}902{,}008{,}176{,}640{,}000 = 20!.$$

16. How many zeros are there at the end of the base 10 representation of 132! ? at the end of the base 2 representation?

17. Does $\left[\frac{x}{n}\right] = \left[\frac{[x]}{n}\right]$ for each real x and each integer $n > 1$?

18. Prove that $0 \leq [nx] - n[x] \leq n-1$ for each real number x and each integer $n \geq 1$.

8–2 TCHEBYCHEV'S THEOREM

In order to establish Tchebychev's result (8–0–2), we must examine some elementary properties of the function $x/\log x$.

Theorem 8–7: *If $f(x) = x/\log x$, then*

$$f(x) \text{ is increasing for } x > e, \tag{8-2-1}$$

$$f(x-2) > \frac{1}{2} f(x) \text{ for } x \geq 4, \tag{8-2-2}$$

$$f\left(\frac{x+2}{2}\right) < \frac{15}{16} f(x) \text{ for } x \geq 8. \tag{8-2-3}$$

PROOF: Since

$$f'(x) = \frac{\log x - 1}{(\log x)^2},$$

we see that $f'(x) > 0$ for $x > e$; this establishes (8-2-1).

To prove (8-2-2), we note that if $x \geq 4$, then $x - 2 \geq x/2$; thus,

$$f(x-2) = \frac{x-2}{\log(x-2)} \geq \frac{x}{2\log(x-2)} > \frac{x}{2\log x} = \frac{1}{2} f(x).$$

To prove (8-2-3), we note that if $x \geq 8$, then $x/2 \geq x^{2/3}$, and $x + 2 \leq 5x/4$; hence,

$$f\left(\frac{x+2}{2}\right) < \frac{x+2}{2\log(x/2)} \leq \frac{x+2}{2\log x^{2/3}} \leq \frac{5x/4}{\frac{4\log x}{3}} = \frac{15}{16} f(x). \qquad \blacksquare$$

We are now ready to prove Tchebychev's theorem.

THEOREM 8-8: *For $x \geq 8$,*

$$\frac{\log 2}{4} \cdot \frac{x}{\log x} < \pi(x) < 30(\log 2) \frac{x}{\log x}.$$

PROOF: Let us examine the binomial coefficient, $\binom{2n}{n}$, the number of combinations of $2n$ things taken n at a time. We recall from Theorem 3-2 that

$$\binom{2n}{n} = \frac{(2n!)}{(n!)(n!)} = \frac{2n(2n-1)\ldots(n+1)}{n(n-1)\ldots 1}.$$

Now, any prime p in the interval $(n, 2n]$ must appear as a factor in the numerator of $\binom{2n}{n}$; since it cannot appear in the denominator (it is larger than n), we see that $p \,\Big|\, \binom{2n}{n}$. Hence, multiplying all such primes together, we find that

$$P_n \,\Big|\, \binom{2n}{n},$$

where P_n denotes the product of all primes larger than n but not exceeding $2n$.

Thus, since each prime appearing as a factor of P_n is larger than n, and since there are $\pi(2n) - \pi(n)$ prime factors of P_n, we see that

$$n^{\pi(2n)-\pi(n)} < P_n < \binom{2n}{n}. \tag{8-2-4}$$

On the other hand, suppose that corresponding to each prime p we define r_p by the inequalities $p^{r_p} \leq 2n < p^{r_p+1}$. Using Theorem 8–6 to determine what power of p appears in the prime factorization of $\binom{2n}{n}$, we see that the correct exponent is the power of p appearing in $(2n)!$ minus the power of p appearing in $(n!)(n!)$; in other words, the exponent is $\sum_{j=1}^{r_p}\left(\left[\frac{2n}{p^j}\right] - 2\left[\frac{n}{p^j}\right]\right)$. By Theorem 8–5,

$$0 \leq \sum_{j=1}^{r_p}\left(\left[\frac{2n}{p^j}\right] - 2\left[\frac{n}{p^j}\right]\right) \leq \sum_{j=1}^{r_p} 1 = r_p.$$

Hence, we see that

$$\binom{2n}{n} \Big| Q_n,$$

where Q_n denotes the product of all p^{r_p}. Since each p^{r_p} does not exceed $2n$, and since Q_n has $\pi(2n)$ factors of the form p^{r_p},

$$\binom{2n}{n} \leq Q_n \leq (2n)^{\pi(2n)}. \tag{8-2-5}$$

As soon as we determine the size of $\binom{2n}{n}$, we shall see that Tchebychev's theorem may be deduced from (8–2–4) and (8–2–5).

By the binomial theorem (Exercise 10 of Section 3–1),

$$(1+x)^{2n} = 1 + \binom{2n}{1}x + \binom{2n}{2}x^2 + \ldots + \binom{2n}{n}x^n + \ldots + x^{2n}.$$

Hence, with $x = 1$, we find that

$$2^{2n} = 1 + \binom{2n}{1} + \binom{2n}{2} + \ldots + \binom{2n}{n} + \ldots + 1 > \binom{2n}{n}. \tag{8-2-6}$$

On the other hand,

$$\binom{2n}{n} = \frac{2n}{n}\,\frac{(2n-1)(2n-2)}{(n-1)(n-2)} \cdots \frac{(n+1)}{1}$$

$$\geq 2 \cdot 2 \cdot 2 \ldots 2 = 2^n. \tag{8-2-7}$$

Combining (8–2–5) with (8–2–7), we find that

$$2^n \le (2n)^{\pi(2n)}. \tag{8-2-8}$$

Taking logarithms of both sides of (8–2–8), we obtain the inequality

$$n \log 2 \le \pi(2n) \log 2n. \tag{8-2-9}$$

Thus, if $x \ge 5$ and $f(x) = x/\log x$, then by (8–2–9), (8–2–1), and (8–2–2),

$$\begin{aligned}\pi(x) &\ge \pi\left(2\left[\frac{x}{2}\right]\right) \ge \frac{\log 2}{2} \cdot \frac{2\left[\frac{x}{2}\right]}{\log 2\left[\frac{x}{2}\right]} \\ &= \frac{\log 2}{2} f\left(2\left[\frac{x}{2}\right]\right) > \frac{\log 2}{2} f(x-2) > \frac{\log 2}{4} f(x) \qquad (8\text{-}2\text{-}10) \\ &= \frac{\log 2}{4} \cdot \frac{x}{\log x},\end{aligned}$$

and this gives the left-hand inequality of Theorem 8–8.

To obtain the other half of Tchebychev's theorem, we combine (8–2–4) with (8–2–6). Thus,

$$n^{\pi(2n)-\pi(n)} < 2^{2n}. \tag{8-2-11}$$

Taking logarithms of both sides of (8–2–11), we find that

$$(\pi(2n) - \pi(n)) \log n < 2n \log 2.$$

Hence,

$$\pi(2n) < (2 \log 2) \frac{n}{\log n} + \pi(n). \tag{8-2-12}$$

We may now establish by mathematical induction that

$$\pi(2n) < 32 (\log 2) \frac{n}{\log n}, \text{ for } n > 1. \tag{8-2-13}$$

First, we note that (8–2–13) is true for $2 \le n \le 8$:

$$\begin{aligned}\pi(4) = 2 < \pi(6) = 3 < \pi(8) = 4 = \pi(10) = 4 \\ < \pi(12) = 5 < \pi(14) = 6 = \pi(16) = 6 < 64 \\ = 32 (\log 2) \frac{2}{\log 2}.\end{aligned}$$

Now assume (8–2–13) for all integers $n \le k$, where $k \ge 8$; then, by (8–2–12) with $f(x) = x/\log x$,

$$\begin{aligned}\pi(2k+2) &< 2(\log 2) f(k+1) + \pi(k+1)\\ &\leqq 2(\log 2) f(k+1) + \pi\left(2\left[\frac{k+2}{2}\right]\right)\\ &< 2(\log 2) f(k+1) + 32(\log 2) f\left(\left[\frac{k+2}{2}\right]\right)\\ &\le 2(\log 2) f(k+1) + 32(\log 2) f\left(\frac{k+2}{2}\right)\\ &< 2(\log 2) f(k+1) + 32(\log 2)\frac{15}{16} f(k+1) \text{ (by (8–2–3))}\\ &= 32(\log 2) f(k+1) = 32(\log 2)\frac{k+1}{\log(k+1)}.\end{aligned}$$

Hence,

$$\pi(2n) < 32(\log 2)\frac{n}{\log n} \text{ for all } n > 1.$$

Thus, for each real number $x \ge 8$,

$$\begin{aligned}\pi(x) < \pi\left(2\left[\frac{x}{2}\right]+2\right) &< 32(\log 2) f\left(\left[\frac{x}{2}\right]+1\right)\\ &\le 32(\log 2) f\left(\frac{x+2}{2}\right)\\ &< 32(\log 2)\frac{15}{16} f(x) \text{ (by (8–2–3))}\\ &= 30(\log 2) f(x)\\ &= 30(\log 2)\frac{x}{\log x}.\end{aligned}$$

■

EXERCISES

1. Deduce from Theorem 8–8 that if x is sufficiently large, there exists a prime between x and $125x$.

2. Prove that if $n \ge 4$ and $2n/3 < p \le n$, then $p \nmid \binom{2n}{n}$.

3. From the definition of r_p on page 108, deduce that if $r_p \ge 2$, then $p \le \sqrt{2n}$.

In Exercises 4 and 5, assume that there are no primes p such that $n < p \le 2n$. We denote the product of all primes not exceeding x by R_x.

4. Using the assumption above and the first inequality in (8–2–5), prove that

$$\binom{2n}{n} \le (2n)^{\sqrt{2n}} R_{\frac{2}{3}n}.$$

5. From (8–2–7) and the assumption that $R_x \le 4^x$ for each real x, deduce that the inequality in Exercise 4 is impossible if n is large.

6. Assuming for each real x that $R_x \le 4^x$, prove Bertrand's postulate: there exist primes between n and $2n$ for n sufficiently large.

7. Using Exercise 12 of Section 8–1, prove that

$$\theta(x) < 30\,(\log 2)x.$$

8–3 SOME UNSOLVED PROBLEMS ABOUT PRIMES

Although primes form the multiplicative building blocks of the integers, many seemingly elementary questions about them are yet unanswered.

For example, in a letter to L. Euler (1742), C. F. Goldbach conjectured that: *every even number larger than 2 is the sum of two primes.* It is simple to show that the statement is true in the case of small numbers; for example, $4=2+2$, $6=3+3$, $8=5+3$, $10=7+3=5+5$, $12=7+5$, $14=11+3=7+7$, $16=13+3=11+5, \ldots$. However, whether the statement is true for *all* even integers is still unsettled. Nevertheless, it is supported by existing evidence. A Russian mathematician, I. M. Vinogradov, proved that all large *odd* integers are the sum of three primes. Surprisingly, his techniques involve extremely subtle use of the theory of complex variables; no one has been able to extend them in order to solve Goldbach's conjecture.

Also unsolved is the famous Twin Primes Problem: *are there infinitely many primes p such that $p+2$ is also a prime?* Thus 3 and 5, 5 and 7, 11 and 13, 17 and 19 are all examples of twin primes. Numerical evidence makes it plausible that infinitely many such pairs exist.

Finally, we mention the *Mersenne Primes*; that is, prime numbers of the form $2^p - 1$ where p is also a prime. We have already met such numbers in Section 3–5 in connection with perfect numbers. M. Mersenne asserted in 1644 that $2^p - 1$ is prime for $p = 2, 3, 5, 7, 13, 17,$ 19, 31, 67, 127, 257, and for no other primes $p < 258$. Actually, $2^{67} - 1$ and $2^{257} - 1$ are not primes while $2^{61} - 1$, $2^{89} - 1$, and $2^{107} - 1$ are; however, it is quite surprising that Mersenne—almost 300 years before the invention of the modern electronic computer—had only five mistakes in his list. In 1963, D. Gillies showed that the primes p not exceeding 12143 for which $2^p - 1$ is also prime are 2, 3, 5, 7, 13, 17, 19, 31, 61, 89, 107, 127, 521, 607, 1279, 2203, 2281, 3217, 4253, 4423, 9689, 9941, and 11213. Recently B. Tuckerman proved that $2^p - 1$ is prime for $p = 19937$. It is not known whether there exist infinitely many Mersenne primes.

Some other unsolved problems about primes, such as the Riemann hypothesis, require considerable background even for the comprehension of their statements. The three problems we have described are among the best known and easiest to understand.

PART II

QUADRATIC CONGRUENCES

In Part I, we saw the importance of congruences to the study of multiplicative questions, and we completely solved the linear congruence

$$ax \equiv b \pmod{c}.$$

Part II is devoted to a more advanced study of congruences involving quadratic polynomials. Perhaps the simplest of these is

$$x^2 \equiv a \pmod{p}$$

where p is a prime. In examining this congruence, we shall encounter Gauss's celebrated Quadratic Reciprocity Law, and we shall acquire information important to the additive problem: in how many ways can an integer be represented as a sum of two squares? Part III contains an extensive investigation of this question.

As you can see, therefore, our study of quadratic congruences serves to link the multiplicative to the additive aspects of number theory.

CHAPTER 9

QUADRATIC RESIDUES

In our study of congruences, we discussed the circumstances under which

$$ax \equiv b \pmod{c}$$

has solutions. The next simplest congruence is

$$x^2 \equiv b \pmod{n}. \tag{9-0-1}$$

The ability to solve (9–0–1) will in most cases enable us to determine whether a quadratic congruence of the form

$$ax^2 + bx + c \equiv 0 \pmod{d}$$

has solutions.

9–1 EULER'S CRITERION

Our first step is to develop a test for determining whether there exists an integer x such that

$$x^2 \equiv a \pmod{p}, \tag{9-1-1}$$

where p is a prime and g.c.d.$(a,p) = 1$. If $p \nmid a$ and (9–1–1) has a solution, we shall say that a is a *quadratic residue* modulo p.

Example 9–1: Let $p = 7$. Since 1, 4, and 9 are perfect squares not divisible by 7, they are quadratic residues modulo 7. Any integer congruent to one of these squares modulo 7 is also a quadratic residue modulo 7; hence -6, 2, and 11 are all quadratic residues modulo 7. Although 49 is a perfect square, it is not a quadratic residue modulo 7 since $7 \mid 49$.

THEOREM 9–1: *The number a is a quadratic residue* modulo p *if and only if*

$$a^{\frac{p-1}{2}} \equiv 1 \pmod{p}.$$

PROOF: Suppose that a is a quadratic residue modulo p. Let X be any integer such that

$$X^2 \equiv a \pmod{p}.$$

Since $p \nmid a$, we see that $p \nmid X$. Consequently,

$$a^{\frac{p-1}{2}} \equiv (X^2)^{\frac{p-1}{2}} \equiv X^{p-1} \equiv 1 \pmod{p},$$

by Euler's theorem (Theorem 5–2).

On the other hand, suppose that

$$a^{\frac{p-1}{2}} \equiv 1 \pmod{p}.$$

Let g be a primitive root modulo p. Then there exists an integer r such that

$$g^r \equiv a \pmod{p},$$

and so

$$g^{r(p-1)/2} \equiv a^{\frac{p-1}{2}} \equiv 1 \pmod{p}.$$

But, from Theorem 7–2, we see that $p-1 \mid r(p-1)/2$. Thus, $r/2$ must be an integer; that is, $r = 2s$, where s is an integer. Hence, if $x = g^s$, then

$$x^2 \equiv g^{2s} \equiv g^r \equiv a \pmod{p};$$

this establishes Euler's Criterion. ■

This proof furnishes a useful corollary.

COROLLARY 9–1: *Let g be a primitive root* modulo p, *and assume* g.c.d.$(a,p) = 1$. *Let r be any integer such that* $g^r \equiv a \pmod{p}$. *Then r is even if and only if a is a quadratic residue* modulo p.

EXERCISE

1. Use Euler's Criterion to determine whether a is a quadratic residue modulo p in each of the following instances:

(a) $a = 2$, $p = 5$; (c) $a = 3$, $p = 11$;

(b) $a = 4$, $p = 7$; (d) $a = 6$, $p = 13$.

9–2 THE LEGENDRE SYMBOL

We shall now introduce the Legendre symbol $\left(\frac{a}{p}\right)$, which greatly simplifies computations in problems on quadratic residues.

DEFINITION 9–1: *If p is an odd prime, then*

$$\left(\frac{a}{p}\right) = \begin{cases} 1 & \text{if } a \text{ is a quadratic residue modulo } p, \\ 0 & \text{if } p \mid a, \\ -1 & \text{otherwise.} \end{cases}$$

THEOREM 9–2: *If p is an odd prime and a and b are relatively prime to p, then*

$$\left(\frac{a}{p}\right) = \left(\frac{b}{p}\right), \text{ if } a \equiv b \pmod{p}, \tag{9-2-1}$$

$$\left(\frac{ab}{p}\right) = \left(\frac{a}{p}\right)\left(\frac{b}{p}\right), \tag{9-2-2}$$

$$a^{\frac{p-1}{2}} \equiv \left(\frac{a}{p}\right) \pmod{p}. \tag{9-2-3}$$

PROOF: Statement (9–2–1) follows directly from the definition of $\left(\frac{a}{p}\right)$, while (9–2–2) follows from Corollary 9–1.

As for (9–2–3), we see that if a is a quadratic residue modulo p, then Theorem 9–1 implies that

$$a^{\frac{p-1}{2}} \equiv 1 = \left(\frac{a}{p}\right) \pmod{p}.$$

If $p \mid a$, then

$$a^{\frac{p-1}{2}} \equiv 0 = \left(\frac{a}{p}\right) \pmod{p}.$$

Finally, if $p \nmid a$, then

$$a^{\frac{p-1}{2}} \equiv \pm 1 \pmod{p},$$

since

$$a^{p-1} \equiv 1 \pmod{p}.$$

Thus, if g.c.d.$(a,p) = 1$ and a is not a quadratic residue modulo p, we see that

$$a^{\frac{p-1}{2}} \equiv -1 = \left(\frac{a}{p}\right) \pmod{p}. \qquad \blacksquare$$

For the following exercises, we extend the definition of the symbol $\left(\frac{a}{m}\right)$ to include the case where m is any odd number:

If $m = p_1 p_2 \ldots p_r$ where the p_i are odd primes (not necessarily distinct), then

$$\left(\frac{n}{m}\right) = \left(\frac{n}{p_1}\right)\left(\frac{n}{p_2}\right) \cdots \left(\frac{n}{p_r}\right).$$

This extended symbol is called the *Jacobi symbol.*

EXERCISES

1. Prove that if c is odd, then $\left(\frac{ab}{c}\right) = \left(\frac{a}{c}\right)\left(\frac{b}{c}\right)$.

2. Prove that if b and c are odd, then $\left(\frac{a}{bc}\right) = \left(\frac{a}{b}\right)\left(\frac{a}{c}\right)$.

3. Prove that if $a \equiv b \pmod{c}$, where c is odd, then $\left(\frac{a}{c}\right) = \left(\frac{b}{c}\right)$.

9–3 THE QUADRATIC RECIPROCITY LAW

In this section, we shall prove a famous theorem of Gauss, the Quadratic Reciprocity Law, which enables us to solve almost all quadratic congruences. This is not an easy theorem; great mathematicians such as L. Euler and A. Legendre were baffled by it. It is some measure of the greatness of C. F. Gauss that he proved it when he was only nineteen years old. Because this theorem is of fundamental importance in number theory, Gauss returned to study it many times throughout his life, and he gave at least six different proofs.*

*That this theorem has occupied the attention of many mathematicians is amply borne out (although slightly exaggerated) in the title of an article by M. Gerstenhaber: "The 152nd Proof of the Law of Quadratic Reciprocity," American Mathematical Monthly, 70(1963), 397–398.

Our first goal, however, is to prove a preliminary theorem called Gauss's Lemma.

Definition 9-2: *If n is any integer, then the* **least residue** *of n modulo m is the integer x in the interval $(-m/2, m/2]$ such that $n \equiv x \pmod{m}$. We denote the least residue of n* modulo *m by $LR_m(n)$.*

Example 9-2: The set $\{-5,-4,-3,-2,-1,0,1,2,3,4,5\}$ is a complete set of least residues modulo 11. Thus $LR_{11}(21) = -1$, since $21 \equiv -1 \pmod{11}$; similarly, $LR_{11}(99) = 0$, and $LR_{11}(60) = 5$.

Definition 9-3: *We define* $\operatorname{sgn}(x)$ *(read signum of x) by*

$$\operatorname{sgn}(x) = \begin{cases} +1 & \text{if } x > 0, \\ 0 & \text{if } x = 0, \\ -1 & \text{if } x < 0. \end{cases}$$

In general, we note that $x = |x| \operatorname{sgn}(x)$.

Theorem 9-3 *(Gauss's Lemma): Let* g.c.d.$(m,p) = 1$ *where p is an odd prime, and let μ be the number of integers in the set*

$$\left\{m, 2m, \ldots, \frac{1}{2}(p-1)m\right\}$$

whose least residues modulo *p are negative. Then*

$$\left(\frac{m}{p}\right) = (-1)^{\mu}.$$

Proof: First note that none of the integers $m, 2m, \ldots, \frac{1}{2}(p-1)m$ is divisible by p. Now, for any n,

$$nm \equiv LR_p(nm) = \operatorname{sgn}(LR_p(nm))\,|LR_p(nm)| \pmod{p}.$$

Since $0 < |LR_p(nm)| < p/2$, we see that as n takes all integral values in $(0,p/2)$, so does $|LR_p(nm)|$. Consequently,

$$\begin{aligned} m(2m)(3m)\ldots((p-1)m/2) \equiv {} & \operatorname{sgn}(LR_p(m))\operatorname{sgn}(LR_p(2m)) \\ & \ldots \operatorname{sgn}(LR_p((p-1)m/2)) \cdot |LR_p(m)| \cdot |LR_p(2m)| \\ & \ldots |LR_p((p-1)m/2)| \pmod{p}, \end{aligned}$$

or

$$\left(\frac{1}{2}(p-1)\right)!\, m^{\frac{p-1}{2}} \equiv \operatorname{sgn}(LR_p(m))\, \operatorname{sgn}(LR_p(2m))$$
$$\ldots \operatorname{sgn}\left(LR_p\left(\frac{1}{2}(p-1)m\right)\right) \cdot \left(\frac{1}{2}(p-1)\right)! \pmod{p}. \quad \text{(9-3-1)}$$

Since $\left(\frac{1}{2}(p-1)\right)!$ is relatively prime to p, we may cancel it from both sides of (9–3–1), by Theorem 4–3. Hence,

$$m^{\frac{p-1}{2}} \equiv \operatorname{sgn}(LR_p(m))\, \operatorname{sgn}(LR_p(2m)) \ldots \operatorname{sgn}\left(LR_p\left(\frac{1}{2}(p-1)m\right)\right) \pmod{p}.$$

But

$$m^{\frac{p-1}{2}} \equiv \left(\frac{m}{p}\right) \pmod{p},$$

by Theorem 9–2, and

$$(-1)^{\mu} = \operatorname{sgn}(LR_p(m))\, \operatorname{sgn}(LR_p(2m)) \ldots \operatorname{sgn}\left(LR_p\left(\frac{1}{2}(p-1)m\right)\right).$$

Thus,

$$\left(\frac{m}{p}\right) \equiv (-1)^{\mu} \pmod{p}.$$

Since each side of this congruence equals either $+1$ or -1, and since $p > 2$,

$$\left(\frac{m}{p}\right) = (-1)^{\mu}. \qquad \blacksquare$$

We are now ready to prove Gauss's Quadratic Reciprocity Law.

THEOREM 9–4 *(Quadratic Reciprocity Law): If p and q are distinct odd primes, then $\left(\frac{p}{q}\right) = \left(\frac{q}{p}\right)$ unless $p \equiv q \equiv 3 \pmod{4}$, in which case $\left(\frac{p}{q}\right) \neq \left(\frac{q}{p}\right)$.*

REMARK: This theorem may be stated equivalently in the form

$$\left(\frac{p}{q}\right)\left(\frac{q}{p}\right) = (-1)^{(p-1)(q-1)/4}.$$

PROOF: Let μ_1 denote the number of integers in the set

$$\left\{q, 2q, \ldots, \frac{1}{2}(p-1)q\right\}$$

with negative least residues modulo p. Let μ_2 denote the number of integers in the set

$$\left\{p, 2p, \ldots, \frac{1}{2}(q-1)p\right\}$$

with negative least residues modulo q. Gauss's Lemma implies that $\left(\frac{p}{q}\right) = (-1)^{\mu_2}$ and $\left(\frac{q}{p}\right) = (-1)^{\mu_1}$. Then, since $\left(\frac{p}{q}\right) = \left(\frac{q}{p}\right)$ if and only if $\left(\frac{p}{q}\right)\left(\frac{q}{p}\right) = 1$, we see that $\left(\frac{p}{q}\right) = \left(\frac{q}{p}\right)$ if and only if

$$(-1)^{\mu_1+\mu_2} = 1.$$

Thus, to prove our theorem, we must show that $\mu_1 + \mu_2$ is odd if and only if $p \equiv q \equiv 3 \pmod 4$.

We proceed geometrically. We shall count in two ways the lattice points (that is, the points whose coordinates are integers) inside a certain hexagon. The first count will show that there are an odd number of lattice points in the hexagon if and only if $p \equiv q \equiv 3 \pmod 4$. The second count will show that there are $\mu_1 + \mu_2$ lattice points inside the hexagon. The two counts together will show that $\mu_1 + \mu_2$ is odd if and only if $p \equiv q \equiv 3 \pmod 4$.

In the first quadrant of the (x,y)-plane, we consider the hexagon H with vertices $ABCDEF$ that lie on the rectangle $AGDJ$ bounded by the coordinate axes and the lines $x = p/2$ and $y = q/2$, as shown in Figure 9–1.

$\overline{EF}$ is defined by

$$y = \frac{q}{p}x + \frac{1}{2},$$

and $\overline{BC}$ is defined by

$$x = \frac{p}{q}y + \frac{1}{2}.$$

The line $\overline{BC}$ has x-intercept $\left(\frac{1}{2}, 0\right)$, and $\overline{EF}$ has y-intercept $\left(0, \frac{1}{2}\right)$; both are parallel to the diagonal $\overline{AD}$.

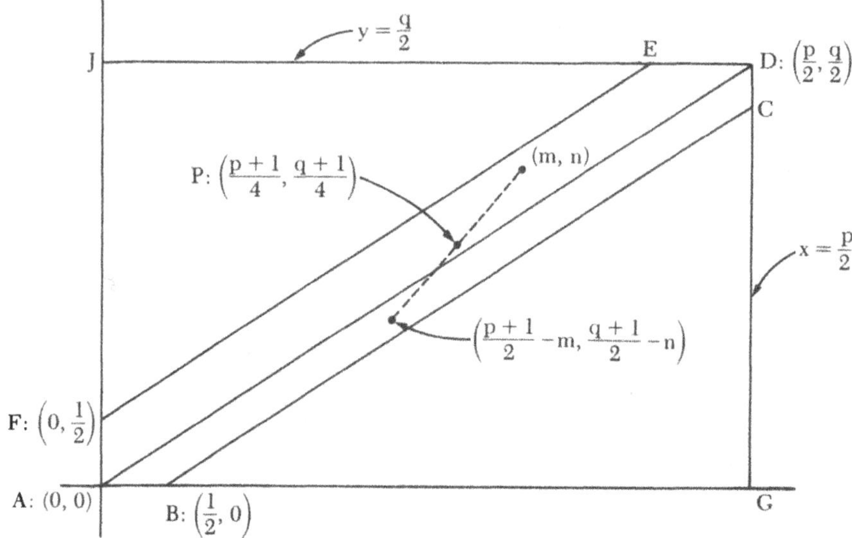

Figure 9–1 Lattice points in proof of Theorem 9–4.

The coordinates of the points* (x,y) that lie in the interior of H must satisfy the inequalities

$$0 < x < p/2, \qquad 0 < y < q/2,$$
$$y < \frac{q}{p}x + \frac{1}{2}, \qquad y > \frac{q}{p}x - \frac{q}{2p}. \tag{9-3-2}$$

Note that if (m,n) is any lattice point in the interior of H, then so is $\left(\frac{p+1}{2} - m, \frac{q+1}{2} - n\right)$; we can verify this by substituting these coordinates into the four inequalities of (9–3–2). Now we see that

$$(m,n) = \left(\frac{p+1}{2} - m, \frac{q+1}{2} - n\right)$$

if and only if $m = \frac{p+1}{4}$ and $n = \frac{q+1}{4}$; and P: $\left(\frac{p+1}{4}, \frac{q+1}{4}\right)$ is a lattice point if and only if $p \equiv q \equiv 3 \pmod 4$. Thus the pairing of (m,n) with $\left(\frac{p+1}{2} - m, \frac{q+1}{2} - n\right)$ shows that the number of lattice points in H is odd if and only if $p \equiv q \equiv 3 \pmod 4$.

*The reader is reminded here that (x,y) is a point and is *not* to be confused with g.c.d.(x,y).

Now let us consider the lattice points in H from a different point of view. It is clear that there are no lattice points on the diagonal $y = \frac{q}{p}x$. If (m,n) is a lattice point below the diagonal, then

$$\frac{qm}{p} - \frac{q}{2p} < n < \frac{qm}{p};$$

that is,

$$-\frac{q}{2} < np - qm < 0. \qquad \text{(9-3-3)}$$

We see that in this case np has a negative least residue modulo q, namely $np - qm$. Conversely, if np has a negative least residue, we can produce an m such that (9–3–3) holds; hence, (m,n) is in H and lies below the diagonal. Thus, since μ_2 is the number of integers in the set $\left\{p, 2p, \ldots, \frac{1}{2}(q-1)p\right\}$ with negative least residues modulo q, there are μ_2 lattice points in H below the diagonal $\overline{AD}$.

In exactly the same way, we find μ_1 lattice points in H above the main diagonal.

We conclude that $\mu_1 + \mu_2$ (the total number of lattice points in H) is odd if and only if $p \equiv q \equiv 3 \pmod 4$; thus the Quadratic Reciprocity Law is proved. ■

To complete our tools for solving quadratic congruences, we need the following consequence of Gauss's Lemma (Theorem 9–3).

THEOREM 9–5: *If p is an odd prime, then*

$$\left(\frac{-1}{p}\right) = (-1)^{\frac{p-1}{2}} \text{ and} \qquad \text{(9-3-4)}$$

$$\left(\frac{2}{p}\right) = (-1)^{\frac{p^2-1}{8}}. \qquad \text{(9-3-5)}$$

PROOF: By Gauss's Lemma, with $m = -1$, we see that $\mu = \frac{1}{2}(p-1)$; this establishes (9–3–4).

With $m = 2$, μ is the number of integers in the set $2, 4, \ldots, p-1$ whose least residues modulo p are negative, which is the same as the number of even integers in the interval $\left[\frac{p+1}{2}, p-1\right]$. Thus,

$$\mu = \begin{cases} 2s & \text{when } p = 8s+1, \\ 2s+1 & \text{when } p = 8s+3, \\ 2s+1 & \text{when } p = 8s+5, \\ 2s+2 & \text{when } p = 8s+7. \end{cases}$$

Hence, μ is even if and only if $p \equiv \pm 1 \pmod 8$. Consequently,

$$\left(\frac{2}{p}\right) = \begin{cases} 1 & \text{when } p \equiv \pm 1 \pmod 8, \\ -1 & \text{when } p \equiv \pm 3 \pmod 8. \end{cases}$$

Now, since

$$(-1)^{\frac{p^2-1}{8}} = \begin{cases} 1 & \text{when } p \equiv \pm 1 \pmod 8, \\ -1 & \text{when } p \equiv \pm 3 \pmod 8, \end{cases}$$

we see that (9–3–5) is established. ■

EXERCISES (In Exercises 1 through 3, the symbol appearing is the Jacobi symbol, defined prior to the exercises for Section 9–2).

1. Prove that if c is odd, then $\left(\frac{-1}{c}\right) = (-1)^{\frac{1}{2}(c-1)}$.

2. Prove that if c is odd, then $\left(\frac{2}{c}\right) = (-1)^{(c^2-1)/8}$.

3. Prove that if a and c are odd and relatively prime, then $\left(\frac{a}{c}\right)\left(\frac{c}{a}\right) = (-1)^{\frac{1}{4}(a-1)(c-1)}$.

4. Use Gauss's Lemma to show that 17 is a quadratic residue modulo 19.

5. Using the Quadratic Reciprocity Law, prove that

$$\left(\frac{3}{p}\right) = \begin{cases} 1 & \text{if } p \equiv 1 \text{ or } 11 \pmod{12}, \\ -1 & \text{if } p \equiv 5 \text{ or } 7 \pmod{12} \end{cases}$$

for each odd prime p.

6. Is it possible that $\left(\frac{n}{m}\right) = 1$ while the congruence $x^2 \equiv n \pmod m$ has no solution? Prove your answer.

7. Does the congruence $x^2 \equiv 631 \pmod{1093}$ have any solutions? [Hint: Use the Jacobi symbol.]

9-4 APPLICATIONS OF THE QUADRATIC RECIPROCITY LAW

We shall limit this discussion to solutions of quadratic congruences with odd moduli. In the following theorem, we consider such congruences for prime power moduli.

THEOREM 9-6: *If p is an odd prime and* g.c.d.$(a,p) = 1$, *then the congruence*

$$x^2 \equiv a \pmod{p^n} \tag{9-4-1}$$

has a solution if and only if $\left(\dfrac{a}{p}\right) = 1$.

PROOF: First note that if $\left(\dfrac{a}{p}\right) = -1$, then the congruence (9-4-1) cannot possibly have solutions, because solutions of (9-4-1) automatically satisfy the condition

$$x^2 \equiv a \pmod{p}.$$

Suppose that $\left(\dfrac{a}{p}\right) = 1$. We establish solutions for (9-4-1) by mathematical induction. Since $\left(\dfrac{a}{p}\right) = 1$, (9-4-1) has solutions when $n = 1$.

Assume now that (9-4-1) has a solution x_0 for $n = k$. Then, there exists an integer m such that

$$x_0^2 - a = mp^k.$$

If $\tilde{x}_0$ is an inverse of x_0 modulo p, then $x_0\tilde{x}_0 = 1 + rp$. Hence,

$$\begin{aligned}
&\left(x_0 - m\tilde{x}_0 \frac{1}{2}(p+1)p^k\right)^2 - a \\
&= x_0^2 - m(p+1)p^k x_0\tilde{x}_0 + m^2\tilde{x}_0^2\left(\frac{1}{2}(p+1)\right)^2 p^{2k} - a \\
&= x_0^2 - m(p+1)p^k(1+rp) + m^2\tilde{x}_0^2\left(\frac{1}{2}(p+1)\right)^2 p^{2k} - a \\
&= -mp^{k+1} + m^2\tilde{x}_0^2\left(\frac{1}{2}(p+1)\right)^2 p^{2k} - mr(p+1)p^{k+1} \\
&= p^{k+1}\left(-m - mr(p+1) + m^2\tilde{x}_0^2\left(\frac{1}{2}(p+1)\right)^2 p^{k-1}\right).
\end{aligned}$$

Therefore, (9–4–1) has a solution for $n = k + 1$, and this establishes our theorem. ■

Let us study specific congruences: We begin with

$$x^2 \equiv 15 \pmod{89}. \tag{9-4-2}$$

Now,

$$\begin{aligned}\left(\frac{15}{89}\right) &= \left(\frac{3}{89}\right)\left(\frac{5}{89}\right) \text{ (by (9–2–2))}\\ &= \left(\frac{89}{3}\right)\left(\frac{89}{5}\right) \text{ (by Theorem 9–4)}\\ &= \left(\frac{2}{3}\right)\left(\frac{4}{5}\right) \text{ (by (9–2–1))}\\ &= (-1)1 = -1.\end{aligned}$$

Thus (9–4–2) has no solutions.

Next we consider the congruence

$$x^2 \equiv 12 \pmod{2989}. \tag{9-4-3}$$

Since $2989 = 7^2 \cdot 61$, where both 7 and 61 are primes, the Chinese Remainder Theorem and Theorem 9–6 tell us that (9–4–3) has solutions if and only if

$$x^2 \equiv 12 \pmod{7} \tag{9-4-4}$$

and

$$x^2 \equiv 12 \pmod{61} \tag{9-4-5}$$

have solutions. Now

$$\left(\frac{12}{61}\right) = \left(\frac{4}{61}\right)\left(\frac{3}{61}\right) = \left(\frac{3}{61}\right) = \left(\frac{61}{3}\right) = \left(\frac{1}{3}\right) = 1$$

and

$$\left(\frac{12}{7}\right) = \left(\frac{4}{7}\right)\left(\frac{3}{7}\right) = \left(\frac{3}{7}\right) = -\left(\frac{7}{3}\right) = -\left(\frac{1}{3}\right) = -1.$$

Hence, (9–4–4) has no solutions, and therefore (9–4–3) has no solutions.

EXERCISES

1. Does $x^2 \equiv 17 \pmod{29}$ have a solution?

2. Does $3x^2 \equiv 12 \pmod{23}$ have a solution?

3. Does $2x^2 \equiv 27 \pmod{41}$ have a solution?

4. Does $x^2 + 5x \equiv 12 \pmod{31}$ have a solution? [Hint: Complete the square.]

5. Does $x^2 \equiv 19 \pmod{30}$ have solutions? [Hint: Use the Chinese Remainder Theorem.]

CHAPTER 10

DISTRIBUTION OF QUADRATIC RESIDUES

This chapter introduces a general counting technique in number theory from which we shall derive sharp and surprising results, with by-products that will be valuable in Chapter 11.

10–1 CONSECUTIVE RESIDUES AND NONRESIDUES

Corollary 9–1 implies that among the set of integers in the closed interval $[1, p-1]$ (where p is an odd prime) there are $(p-1)/2$ quadratic residues and $(p-1)/2$ nonresidues modulo p. Now, how many *consecutive pairs* of quadratic residues are there?

If we can argue that there is equal likelihood for a particular integer to be a quadratic residue or a nonresidue, approximately one-fourth of the consecutive integer pairs in $[1, p-1]$ should be consecutive quadratic residues. We shall see that our estimate is correct.

Theorem 10–1: $N(p) = \frac{1}{4}\left(p - 4 - (-1)^{(p-1)/2}\right)$, *where* $N(p)$ *denotes the number of pairs of consecutive quadratic residues* modulo p *in* $[1, p-1]$.

Proof: Let $c_p(n)$ be defined by the formula

$$c_p(n) = \begin{cases} 1 & \text{if both } n \text{ and } n+1 \text{ are quadratic residues modulo } p, \\ 0 & \text{otherwise.} \end{cases}$$

Then,

$$N(p) = \sum_{n=1}^{p-2} c_p(n). \tag{10-1-1}$$

We note that

$$c_p(n) = \frac{1}{4}\left(1 + \left(\frac{n}{p}\right)\right)\left(1 + \left(\frac{n+1}{p}\right)\right), \tag{10-1-2}$$

since the right side of (10–1–2) is 0 if either n or $n+1$ is a nonresidue, and 1 when both are quadratic residues. Hence,

$$N(p) = \frac{1}{4}\sum_{n=1}^{p-2}\left(1 + \left(\frac{n}{p}\right)\right)\left(1 + \left(\frac{n+1}{p}\right)\right)$$

$$= \frac{1}{4}\sum_{n=1}^{p-2}\left(1 + \left(\frac{n}{p}\right) + \left(\frac{n+1}{p}\right) + \left(\frac{n}{p}\right)\left(\frac{n+1}{p}\right)\right)$$

$$= \frac{1}{4}\sum_{n=1}^{p-2} 1 + \frac{1}{4}\sum_{n=1}^{p-2}\left(\frac{n}{p}\right) + \frac{1}{4}\sum_{n=1}^{p-2}\left(\frac{n+1}{p}\right) + \frac{1}{4}\sum_{n=1}^{p-2}\left(\frac{n}{p}\right)\left(\frac{n+1}{p}\right). \tag{10-1-3}$$

The first sum in (10–1–3) is equal to $(p-2)/4$. The second and third sums are relatively easy to evaluate. Since there are as many quadratic residues as nonresidues modulo p, we see that $\left(\frac{n}{p}\right)$ is equally often -1 and $+1$ in the interval $[1, p-1]$. Consequently,

$$\sum_{n=1}^{p-1}\left(\frac{n}{p}\right) = 0, \tag{10-1-4}$$

and therefore

$$\sum_{n=1}^{p-2}\left(\frac{n}{p}\right) = -\left(\frac{p-1}{p}\right) = -\left(\frac{-1}{p}\right) = -(-1)^{(p-1)/2},$$

and

$$\sum_{n=1}^{p-2}\left(\frac{n+1}{p}\right) = -\left(\frac{1}{p}\right) = -1.$$

If we could also prove that

$$\sum_{n=1}^{p-2}\left(\frac{n}{p}\right)\left(\frac{n+1}{p}\right) = -1, \tag{10-1-5}$$

then we would have the formula

$$N(p) = \frac{1}{4}\,(p - 2 - (-1)^{(p-1)/2} - 1 - 1)$$

$$= \frac{1}{4}\,(p - 4 - (-1)^{(p-1)/2}).$$

Theorem 10–3 will complete this proof by establishing (10–1–5). ■

Since sums similar to (10–1–5) will also play a significant role in Theorem 10–4, we shall prove two general, related results.

THEOREM 10–2: *Suppose c_j is defined for all integers j, and $c_j = c_k$ whenever $j \equiv k \pmod{n}$. Let $r_1, r_2, \ldots, r_n$ be any complete residue system* modulo n. *Then*

$$\sum_{j=0}^{n-1} c_j = c_{r_1} + c_{r_2} + \ldots + c_{r_n}. \tag{10-1-6}$$

REMARK: We often abbreviate (10–1–6) by writing

$$\sum_{j=0}^{n-1} c_j = \sum_{r \pmod{n}} c_r,$$

where the second sum means $c_{r_1} + c_{r_2} + \ldots + c_{r_n}$.

PROOF: Since both $\{r_1, r_2, \ldots, r_n\}$ and $\{0, 1, 2, \ldots, n-1\}$ constitute complete residue systems, each nonnegative integer i less than n is congruent to precisely one r_j; thus $c_i = c_{r_j}$. Hence, the sums $c_0 + c_1 + \ldots + c_{n-1}$ and $c_{r_1} + c_{r_2} + \ldots + c_{r_n}$ are merely commutations of one another and are therefore equal. ■

THEOREM 10–3: *If p is an odd prime, then*

$$\sum_{n=0}^{p-1} \left(\frac{(n-a)(n-b)}{p}\right) = \begin{cases} p-1 & \text{if } a \equiv b \pmod{p}, \\ -1 & \text{if } a \not\equiv b \pmod{p}. \end{cases}$$

REMARK: (10–1–5) is the case $a = 0$, $b = -1$.

PROOF: By Theorem 10–2,

$$\sum_{n=0}^{p-1} \left(\frac{(n-a)(n-b)}{p}\right) = \sum_{n \pmod{p}} \left(\frac{(n-a)(n-b)}{p}\right).$$

Now, as n assumes all values in a complete residue system modulo p, so does $n + a$. Thus

$$\sum_{n \pmod{p}} \left(\frac{(n-a)(n-b)}{p}\right) = \sum_{n \pmod{p}} \left(\frac{n(n+a-b)}{p}\right).$$

If $a \equiv b \pmod{p}$, then

$$\left(\frac{n(n+a-b)}{p}\right) = \left(\frac{n^2}{p}\right) = 1, \text{ for } n \not\equiv 0 \pmod{p},$$

and thus

$$\sum_{n=0}^{p-1}\left(\frac{(n-a)(n-b)}{p}\right)=p-1.$$

If $a \not\equiv b \pmod p$, let $\lambda = a-b$; then $\lambda \not\equiv 0 \pmod p$. Now $\left(\frac{n}{p}\right)=0$, if $n \equiv 0 \pmod p$; consequently, we see that

$$\sum_{n \pmod p}\left(\frac{n(n+\lambda)}{p}\right)=\sum_{\substack{n \pmod p\\ n\not\equiv 0 \pmod p}}\left(\frac{n(n+\lambda)}{p}\right).$$

If $n \not\equiv 0 \pmod p$, there exists an $\tilde{n}$ such that $n\tilde{n} \equiv 1 \pmod p$. Furthermore, $\left(\frac{\tilde{n}^2}{p}\right)=1$. Thus,

$$\begin{aligned}\sum_{\substack{n \pmod p\\ n\not\equiv 0 \pmod p}}\left(\frac{n(n+\lambda)}{p}\right)&=\sum_{\substack{n \pmod p\\ n\not\equiv 0 \pmod p}}\left(\frac{\tilde{n}^2}{p}\right)\left(\frac{n(n+\lambda)}{p}\right)\\&=\sum_{\substack{n \pmod p\\ n\not\equiv 0 \pmod p}}\left(\frac{\tilde{n}n(\tilde{n}n+\lambda\tilde{n})}{p}\right)\\&=\sum_{\substack{n \pmod p\\ n\not\equiv 0 \pmod p}}\left(\frac{1+\lambda\tilde{n}}{p}\right).\end{aligned}$$

But now, as n assumes all values in a reduced residue system modulo p, so does $\tilde{n}$; since $\lambda \not\equiv 0 \pmod p$, so does $\lambda\tilde{n}$. Thus, if we set $m=\lambda\tilde{n}$, then

$$\begin{aligned}\sum_{\substack{n \pmod p\\ n\not\equiv 0 \pmod p}}\left(\frac{n(n+\lambda)}{p}\right)&=\sum_{\substack{m \pmod p\\ m\not\equiv 0 \pmod p}}\left(\frac{1+m}{p}\right)\\&=\sum_{m \pmod p}\left(\frac{1+m}{p}\right)-\left(\frac{1}{p}\right)\\&=0-1 \text{ (by (10–1–4))}\\&=-1.\end{aligned}$$

■

We have now also proved Theorem 10–1, since Theorem 10–3 establishes (10–1–5). With minor alterations, the technique used in Theorem 10–1, combined with the general result in Theorem 10–3, yields similar results for consecutive nonresidues modulo p. Such results are listed in the following exercises.

EXERCISES

1. Let $N_1(p)$ denote the number of pairs of integers in $[1,p-1]$ in which the first is a quadratic residue and the second is a quadratic nonresidue modulo p. Prove that $N_1(p) = \frac{1}{4}\,(p-(-1)^{(p-1)/2})$.

2. Let $N_2(p)$ denote the number of pairs of integers in $[1,p-1]$ in which the first is a quadratic nonresidue and the second is a quadratic residue modulo p. Prove that $N_2(p) = \frac{1}{4}\,(p-2+(-1)^{(p-1)/2})$.

3. Let $N_3(p)$ denote the number of pairs of integers in $[1,p-1]$ in which the first is a quadratic nonresidue and the second is a quadratic nonresidue modulo p. Prove that $N_3(p) = \frac{1}{4}\,(p-2+(-1)^{(p-1(/2)})$.

4. Suppose $Y(f)$ denotes the number of mutually incongruent solutions (x,y) of the congruence $y^2 \equiv f(x) \pmod p$ (p an odd prime), where $f(x)$ is a polynomial with integral coefficients. Prove that

$$Y(f) = p + \sum_{n=0}^{p-1} \left(\frac{f(n)}{p}\right).$$

   Note: Two solutions (x_1,y_1) and (x_2,y_2) are said to be mutually incongruent if either $x_1 \not\equiv x_2 \pmod p$ or $y_1 \not\equiv y_2 \pmod p$, or both.

5. Prove that there are $p-1$ mutually incongruent pairs of solutions of the congruence $y^2 \equiv x^2+3x+2 \pmod p$ (p an odd prime).

6. Suppose g.c.d.$(n,p)=1$, where p is a prime and $n = q_1^{a_1}\ldots q_r^{a_r}$, with q_i prime. Let $D(n)$ denote the number of divisors of n that are square-free and are quadratic residues modulo p. Prove that

$$D(n) = \frac{1}{2}\sum_{d\mid n}\left(1+\left(\frac{d}{p}\right)\right)|\mu(d)|.$$

7. Using the notation of Exercise 6, prove that

$$D(n) = \begin{cases} 2^r & \text{if all } q_i \text{ are quadratic residues modulo } p, \\ 2^{r-1} & \text{otherwise.} \end{cases}$$

10–2 CONSECUTIVE TRIPLES OF QUADRATIC RESIDUES

An approach similar to that in Theorem 10–1 leads to formulae for the number of consecutive triples of quadratic residues modulo p. The results in this section are somewhat harder to prove than those in Section 10–1; however, we not only obtain our main objective, but also unexpectedly discover an important result on numbers representable as a sum of two squares.

Definition 10–1: *Let $\nu(p)$ denote the number of consecutive triples of quadratic residues in the interval $[1,p-1]$.*

Example 10–1: If $p=11$, the quadratic residues less than 11 are 1, 3, 4, 5, and 9; consequently, $\nu(11)=1$. As a check on Theorem 10–1, we note that $N(11)=2=\frac{1}{4}(11-4+1)$.

Theorem 10–4: *If p is an odd prime, then*

$$\nu(p)=\frac{1}{8}p+E_p,$$

where $|E_p|<\frac{1}{4}\sqrt{p}+2$.

Remark: As we shall see in the proof, a much more explicit but cumbersome value for E_p can be given, especially when $p\equiv 3 \pmod 4$.

Proof: Let $C_p(n)$ be defined by the formula

$$C_p(n) = \begin{cases} 1 & \text{if } n,\ n+1, \text{ and } n+2 \text{ are all quadratic residues} \\ & \text{modulo } p, \\ 0 & \text{otherwise.} \end{cases}$$

Then,

$$\nu(p)=\sum_{n=1}^{p-3} C_p(n).$$

As in Theorem 10–1, we see directly that

$$C_p(n) = \frac{1}{8}\left(1 + \left(\frac{n}{p}\right)\right)\left(1 + \left(\frac{n+1}{p}\right)\right)\left(1 + \left(\frac{n+2}{p}\right)\right).$$

Consequently,

$$\begin{aligned}
\nu(p) &= \frac{1}{8}\sum_{n=1}^{p-3}\left(1 + \left(\frac{n}{p}\right)\right)\left(1 + \left(\frac{n+1}{p}\right)\right)\left(1 + \left(\frac{n+2}{p}\right)\right) \\
&= \frac{1}{8}\sum_{n=1}^{p-3}\left(1 + \left(\frac{n}{p}\right) + \left(\frac{n+1}{p}\right) + \left(\frac{n+2}{p}\right) + \left(\frac{n}{p}\right)\left(\frac{n+1}{p}\right)\right. \\
&\qquad + \left(\frac{n}{p}\right)\left(\frac{n+2}{p}\right) + \left(\frac{n+1}{p}\right)\left(\frac{n+2}{p}\right) \\
&\qquad \left. + \left(\frac{n}{p}\right)\left(\frac{n+1}{p}\right)\left(\frac{n+2}{p}\right)\right) \\
&= \frac{1}{8}\sum_{n=1}^{p-3} 1 + \frac{1}{8}\sum_{n=1}^{p-3}\left(\frac{n}{p}\right) + \frac{1}{8}\sum_{n=1}^{p-3}\left(\frac{n+1}{p}\right) + \frac{1}{8}\sum_{n=1}^{p-3}\left(\frac{n+2}{p}\right) \\
&\qquad + \frac{1}{8}\sum_{n=1}^{p-3}\left(\frac{n}{p}\right)\left(\frac{n+1}{p}\right) + \frac{1}{8}\sum_{n=1}^{p-3}\left(\frac{n}{p}\right)\left(\frac{n+2}{p}\right) \\
&\qquad + \frac{1}{8}\sum_{n=1}^{p-3}\left(\frac{n+1}{p}\right)\left(\frac{n+2}{p}\right) \\
&\qquad + \frac{1}{8}\sum_{n=1}^{p-3}\left(\frac{n}{p}\right)\left(\frac{n+1}{p}\right)\left(\frac{n+2}{p}\right).
\end{aligned}$$

Now the first seven of the last eight sums are of the type treated in Theorems 10–1 and 10–3. Therefore,

$$\begin{aligned}
\nu(p) = \frac{1}{8}&\left((p-3) + \left(0 - \left(\frac{p-2}{p}\right) - \left(\frac{p-1}{p}\right)\right) + \left(0 - \left(\frac{1}{p}\right) - \left(\frac{p-1}{p}\right)\right)\right. \\
&+ \left(0 - \left(\frac{1}{p}\right) - \left(\frac{2}{p}\right)\right) + \left(-1 - \left(\frac{(p-2)(p-1)}{p}\right)\right) \\
&+ \left(-1 - \left(\frac{(p-1)(p+1)}{p}\right)\right) + \left(-1 - \left(\frac{2}{p}\right)\right) \\
&\left. + \sum_{n=1}^{p-3}\left(\frac{n(n+1)(n+2)}{p}\right)\right).
\end{aligned} \tag{10-2-1}$$

Hence, if $E_p = \nu(p) - \frac{1}{8}p$, then by (10-2-1)

$$|E_p| < \frac{3}{8} + \frac{1}{4} + \frac{1}{4} + \frac{1}{4} + \frac{1}{4} + \frac{1}{4} + \frac{1}{4} + \frac{1}{8}\left|\sum_{n=1}^{p-3}\left(\frac{n(n+1)(n+2)}{p}\right)\right|$$

$$< 2 + \frac{1}{8}\left|\sum_{n=1}^{p-3}\left(\frac{n(n+1)(n+2)}{p}\right)\right|.$$

To establish Theorem 10-4, we need only show that

$$\left|\sum_{n=1}^{p-3}\left(\frac{n(n+1)(n+2)}{p}\right)\right| \leq 2\sqrt{p}. \tag{10-2-2}$$

Define $S(m)$ by the formula

$$S(m) = \sum_{n(\text{mod } p)}\left(\frac{n(n^2-m)}{p}\right).$$

Then,

$$S(1) = \sum_{n(\text{mod } p)}\left(\frac{n(n-1)(n+1)}{p}\right)$$

$$= \sum_{n(\text{mod } p)}\left(\frac{(n+1)n(n+2)}{p}\right)$$

$$= \sum_{n=1}^{p-3}\left(\frac{n(n+1)(n+2)}{p}\right).$$

Also,

$$S(m) = \sum_{n=1}^{(p-1)/2}\left(\frac{n(n^2-m)}{p}\right) + \sum_{n=(p+1)/2}^{p-1}\left(\frac{n(n^2-m)}{p}\right)$$

$$= \sum_{n=1}^{(p-1)/2}\left(\frac{n(n^2-m)}{p}\right) + \sum_{n=1}^{(p-1)/2}\left(\frac{(p-n)((p-n)^2-m)}{p}\right)$$

$$= \sum_{n=1}^{(p-1)/2}\left(\frac{n(n^2-m)}{p}\right) + \left(\frac{-1}{p}\right)\sum_{n=1}^{(p-1)/2}\left(\frac{n(n^2-m)}{p}\right).$$

If $p \equiv 3 \pmod 4$, then $\left(\frac{-1}{p}\right) = -1$, and so $S(m) = 0$ in this case. Therefore,

$$|E_p| < 2, \text{ if } p \equiv 3 \pmod 4.$$

Now, if $p \equiv 1 \pmod 4$, we may only conclude that $S(m)$ is an even number, since $\left(\frac{-1}{p}\right) = 1$ and

$$S(m) = 2 \sum_{n=1}^{(p-1)/2} \left(\frac{n(n^2 - m)}{p}\right).$$

Suppose k is any integer such that $k \not\equiv 0 \pmod p$. Then $\left(\frac{k^4}{p}\right) = 1$, and

$$\begin{aligned} S(m) &= \sum_{n(\text{mod } p)} \left(\frac{n(n^2 - m)}{p}\right) \\ &= \sum_{n(\text{mod } p)} \left(\frac{k^4}{p}\right)\left(\frac{n(n^2 - m)}{p}\right) \\ &= \sum_{n(\text{mod } p)} \left(\frac{kkn(k^2n^2 - k^2m)}{p}\right) \\ &= \left(\frac{k}{p}\right) \sum_{n(\text{mod } p)} \left(\frac{kn((kn)^2 - k^2m)}{p}\right). \end{aligned}$$

As n assumes all values in a complete residue system modulo p, so does kn. Thus if $h = kn$, we see that

$$\begin{aligned} S(m) &= \left(\frac{k}{p}\right) \sum_{h(\text{mod } p)} \left(\frac{h(h^2 - k^2m)}{p}\right) \\ &= \left(\frac{k}{p}\right) S(k^2m). \end{aligned}$$

If j is a quadratic residue modulo p, then $j \equiv c^2 \pmod p$ for some c, and

$$S(1) = \left(\frac{c}{p}\right) S(c^2) = \left(\frac{c}{p}\right) S(j).$$

Thus

$$|S(1)| = |S(j)|. \tag{10-2-3}$$

Also, if l and r' are any quadratic nonresidues modulo p, then by Corollary 9–1

$$l \equiv g^{2a+1} \pmod p,$$

and

$$r' \equiv g^{2b+1} \pmod{p}.$$

Hence, if $b \geq a$, and if we set $t = g^{b-a}$, then $r' \equiv lt^2 \pmod{p}$, and

$$S(l) = \left(\frac{t}{p}\right) S(lt^2) = \left(\frac{t}{p}\right) S(r').$$

Consequently,

$$|S(l)| = |S(r')|. \tag{10-2-4}$$

Hence we see that the absolute value of $S(m)$ is completely determined by whether m is a quadratic residue modulo p. Consequently, with $p \equiv 1 \pmod 4$,

$$\frac{(p-1)}{2} S(1)^2 + \frac{(p-1)}{2} S(l)^2 = \sum_{m=1}^{p-1} S(m)^2$$

$$= \sum_{m \pmod p} S(m)^2$$

$$= \sum_{m \pmod p} \left(\sum_{s \pmod p} \left(\frac{s(s^2-m)}{p} \right) \right) \left(\sum_{t \pmod p} \left(\frac{t(t^2-m)}{p} \right) \right)$$

$$= \sum_{m \pmod p} \sum_{s \pmod p} \sum_{t \pmod p} \left(\frac{st}{p}\right) \left(\frac{(m-s^2)(m-t^2)}{p} \right)$$

$$= \sum_{s \pmod p} \sum_{t \pmod p} \left(\frac{st}{p}\right) \sum_{m \pmod p} \left(\frac{(m-s^2)(m-t^2)}{p} \right).$$

Now, applying Theorem 10–3 to the inner sum, we obtain the equation

$$\frac{(p-1)}{2} S(1)^2 + \frac{(p-1)}{2} S(l)^2 = \sum_{s \pmod p} \sum_{\substack{t \pmod p \\ t^2 \equiv s^2 \pmod p}} \left(\frac{st}{p}\right)(p-1)$$

$$+ \sum_{s \pmod p} \sum_{\substack{t \pmod p \\ t^2 \not\equiv s^2 \pmod p}} \left(\frac{st}{p}\right)(-1)$$

$$= 2(p-1)(p-1) - \sum_{s \pmod p} \left(\frac{s}{p}\right)\left(0 - \left(\frac{s}{p}\right) - \left(\frac{-s}{p}\right)\right)$$

$$= 2(p-1)^2 + 2(p-1) = 2p(p-1).$$

Hence

$$\left(\frac{S(1)}{2}\right)^2 + \left(\frac{S(l)}{2}\right)^2 = p. \tag{10-2-5}$$

Thus $|S(1)| < 2p^{1/2}$, and this establishes (10–2–2). ■

COROLLARY 10–1: *Every prime* $p \equiv 1 \pmod 4$ *is representable as a sum of two squares.*

PROOF: The assertion follows from (10–2–5). ■

This corollary, which seems almost unrelated to Theorem 10–4, is an important result in itself and will form the basis of the next chapter.

EXERCISES

1. Use the results of Theorem 10–4 and Corollary 10–1 to construct solutions of $x^2 + y^2 = 29$.

2. Using Exercise 4 of Section 10–1, prove that the congruence

$$y^2 \equiv x^3 + 3x^2 + 2x \pmod p$$

has p mutually incongruent solutions if $p \equiv 3 \pmod 4$ and that it has at least $p - 2\sqrt{p}$ mutually incongruent solutions if $p \equiv 1 \pmod 4$.

3. Evaluate $\sum_{n \pmod p} \left(\frac{n(n+1)(n+2)(n+3)}{p}\right)$ for $p = 5, 7, 11,$ and 13. How does the absolute value of your result compare with $3\sqrt{p}$?

PART III

ADDITIVITY

Part II, Quadratic Congruences, has yielded important information about the representation of integers as sums of two squares. In Chapter 11, we shall investigate in detail the subject of representing integers as sums of squares. This problem is part of the broader question: *In how many ways can a nonnegative integer n be represented as a sum of elements of a set S?* Aside from letting S be the set of perfect squares, we might take S to be the set of all positive integers, or the set of all odd positive integers, or any set whose elements are integers. From these possibilities, there arise many surprising results which we shall study in Chapters 12, 13, and 14.

Chapter 12, devoted to simpler additive concepts, suggests ways to make conjectures in additive number theory. Techniques of generating functions are introduced in Chapter 13, and are then used in Chapter 14 to prove some of the conjectures formed in Chapter 12.

CHAPTER 11

SUMS OF SQUARES

Recalling the lagniappe we obtained as a corollary to Theorem 10–4, we shall now find *all* numbers representable as a sum of two or four squares. The problem of representing integers as sums of squares dates back to Diophantus of Alexandria, who apparently knew that every integer is the sum of two, three or four positive integral squares.

EXERCISE

What does "lagniappe" mean, and why does it apply to Corollary 10–1?

11–1 SUMS OF TWO SQUARES

THEOREM 11–1: *A positive integer n can be represented as a sum of two squares if and only if its factorization into powers of distinct primes contains no odd powers of primes congruent to 3 modulo 4.*

PROOF: Suppose n has the prime factorization

$$n = 2^{\alpha} p_1^{\beta_1} p_2^{\beta_2} \ldots p_r^{\beta_r} q_1^{\gamma_1} q_2^{\gamma_2} \ldots q_s^{\gamma_s},$$

where $p_i \equiv 1 \pmod 4$ $(1 \le i \le r)$, and where $q_j \equiv 3 \pmod 4$ $(1 \le j \le s)$. Suppose that at least one of the γ_i is odd, say γ_1. If

$$n = x^2 + y^2$$

and $d = \text{g.c.d.}(x,y)$, then, with $n_1 = n/d^2$, $x_0 = x/d$, and $y_0 = y/d$, we can assert that $\text{g.c.d.}(x_0, y_0) = 1$ and

$$n_1 = n/d^2 = (x/d)^2 + (y/d)^2 = x_0^2 + y_0^2.$$

Now let $\tilde{\gamma}$ be the exponent of q_1 appearing in the prime factorization of n_1; $\tilde{\gamma}$ is odd, because $n_1 = n/d^2$. Since q_1 divides $n_1 = x_0^2 + y_0^2$, q_1 could not divide either x_0 or y_0 without dividing the other; and, since g.c.d.$(x_0, y_0) = 1$, we conclude that g.c.d.$(x_0, q_1) =$ g.c.d.$(y_0, q_1) = 1$. Hence, there exists an integer u such that

$$x_0 \equiv y_0 u \pmod{q_1},$$

by Theorem 5–1. Thus,

$$0 \equiv n_1 = x_0^2 + y_0^2 \equiv u^2 y_0^2 + y_0^2 \equiv y_0^2(1 + u^2) \pmod{q_1}.$$

Since q_1 does not divide y_0^2, we see that

$$u^2 + 1 \equiv 0 \pmod{q_1}.$$

Consequently,

$$u^2 \equiv -1 \pmod{q_1},$$

which, by Theorem 9–5, is impossible. Hence, all the γ_i must be even if n is to be represented as the sum of two squares.

Suppose now that all the γ_i are even, say $\gamma_i = 2\delta_i$. Thus,

$$n = 2^{\alpha} p_1^{\beta_1} \ldots p_r^{\beta_r} (q_1^2)^{\delta_1} \ldots (q_s^2)^{\delta_s}.$$

Now $2 = 1^2 + 1^2$, and $q_i^2 = q_i^2 + 0^2$.* Furthermore, Corollary 10–1 asserts that each p_i is representable as the sum of two squares. Thus n is a product of numbers each of which is a sum of two squares.

Note that

$$(x^2 + y^2)(v^2 + w^2) = (xv + yw)^2 + (xw - yv)^2. \tag{11-1-1}$$

Hence, whenever two numbers are representable as a sum of two squares, so is their product; moreover, mathematical induction extends this assertion to products of arbitrarily many factors. Thus n is indeed representable as a sum of two squares. ∎

Example 11–1: Since $3 \equiv 3 \pmod 4$, $5 \equiv 1 \pmod 4$, and $7 \equiv 3 \pmod 4$, the factorization of $315 = 3^2 \cdot 5 \cdot 7$ contains an odd power of a prime congruent to 3 modulo 4. By Theorem 11–1, 315 cannot be represented as a sum of two squares.

---

*If none of the prime factors of n are congruent to 3 modulo 4, we merely alter our argument by omitting the factors q_i.

Example 11–2: The factorization of $3185 = 5 \cdot 7^2 \cdot 13$ contains no odd power of a prime congruent to 3 modulo 4. Hence 3185 is representable as a sum of two squares. To find two squares that sum to 3185, we first represent 5, 7^2, and 13:

$$5 = 2^2 + 1^2,$$
$$7^2 = 7^2 + 0^2,$$
$$13 = 3^2 + 2^2.$$

Therefore,

$$\begin{aligned} 3185 &= 5 \cdot 7^2 \cdot 13 \\ &= (2^2 + 1^2)(7^2 + 0^2)(3^2 + 2^2) \\ &= (14^2 + 7^2)(3^2 + 2^2) \\ &= (14 \cdot 3 + 7 \cdot 2)^2 + (14 \cdot 2 - 7 \cdot 3)^2 \\ &= 56^2 + 7^2. \end{aligned}$$

EXERCISES

1. Represent 274625 $(= 5^3 \cdot 13^3)$ as the sum of two integral squares.

2. Represent 333 as the sum of two integral squares.

3. We recall that if $z = x + iy$ is a complex number, then $|z| = \sqrt{x^2 + y^2}$ and $\bar{z} = x - iy$.

   (a) Prove that $z\bar{z} = |z|^2$

   (b) Prove that if $w = u + iv$, then

   $$|w|\,|z| = |wz|.$$

   Discuss the relation of this last result to equation (11–1–1).

11–2 SUMS OF FOUR SQUARES

The proof that every positive integer can be represented as a sum of four nonnegative integral squares requires a preliminary result.

THEOREM 11–2: *For each prime p there exist integers A, B, and C, not all zero, such that*

$$A^2 + B^2 + C^2 \equiv 0 \pmod{p}.$$

PROOF: If $p = 2$, take $A = 1$, $B = 1$, $C = 0$. If $p \equiv 1 \pmod 4$, choose A as a solution of $x^2 \equiv -1 \pmod p$, $B = 1$, $C = 0$. If $p \equiv 3 \pmod 4$, set $C = 1$. Then we must solve the congruence

$$A^2 + B^2 \equiv -1 \pmod p.$$

Let d be the least positive nonresidue modulo p; then $\left(\frac{-d}{p}\right) = \left(\frac{-1}{p}\right)\left(\frac{d}{p}\right) = (-1)(-1) = 1$, and therefore $-d$ is a quadratic residue modulo p. Also, $d \geq 2$, since d is a nonresidue. Choose B so that $B^2 \equiv -d \pmod p$. Then we must find an A such that

$$A^2 \equiv d - 1 \pmod p.$$

Note that $d - 1$ is a quadratic residue modulo p, since it is both positive and less than d, the *least* positive nonresidue modulo p. Thus, this last congruence is clearly solvable. ■

We are now prepared to prove the famous result of Lagrange.

THEOREM 11–3: *Every positive integer is a sum of four nonnegative integral squares.*

PROOF: Our theorem on two squares relied on equation (11–1–1), a rather cumbersome algebraic identity. An equation similar to (11–1–1) is also necessary here, namely

$$\begin{aligned}(a^2 + b^2 + c^2 + d^2)(e^2 + f^2 + g^2 + h^2) &= (ae + bf + cg + dh)^2 + (af - be - ch + dg)^2 \\ &+ (ag + bh - ce - df)^2 + (ah - bg + cf - de)^2. \end{aligned} \tag{11-2-1}$$

This equation shows that if each multiplicand of a product is representable as a sum of four squares, so is the product.

To complete our theorem we need only show that every odd prime is representable as a sum of four squares.

By Theorem 10–2, we know that there exist integers A, B, and C, not all zero, such that

$$A^2 + B^2 + C^2 \equiv 0 \pmod p. \tag{11-2-2}$$

We may write (11–2–2) equivalently as

$$A^2 + B^2 + C^2 + D^2 = Kp, \tag{11-2-3}$$

where K is an integer and $D=0$. Take k to be the least positive integer such that kp is the sum of four nonnegative integral squares. Equation (11–2–3) and the least-integer principle (see Exercise 2 of Section 2–1) guarantee that k must exist.

In (11–2–2) and (11–2–3), we may choose A, B, C, and D in the interval $\left[0, \frac{p}{2}\right)$; thus,

$$kp < 4\left(\frac{p}{2}\right)^2 = p^2;$$

that is,

$$k < p.$$

To finish the proof, we need only show that $k = 1$. We therefore assume that $k > 1$, and we derive a contradiction by first considering k odd, and second, k even.

Suppose that $k > 1$ and k is odd. Then with

$$kp = a^2 + b^2 + c^2 + d^2, \tag{11-2-4}$$

we choose $\bar{a}$, $\bar{b}$, $\bar{c}$, and $\bar{d}$ from the interval $\left[0, \frac{k}{2}\right)$ so that $a \equiv \bar{a}$, $b \equiv \bar{b}$, $c \equiv \bar{c}$, and $d \equiv \bar{d} \pmod{k}$. Hence,

$$\bar{a}^2 + \bar{b}^2 + \bar{c}^2 + \bar{d}^2 \equiv 0 \pmod{k},$$

and thus there exists $l \geq 0$ such that

$$kl = \bar{a}^2 + \bar{b}^2 + \bar{c}^2 + \bar{d}^2. \tag{11-2-5}$$

Clearly $l < k$, since each of $\bar{a}$, $\bar{b}$, $\bar{c}$, and $\bar{d}$ is less than $\frac{k}{2}$. Now suppose that l is zero; then $\bar{a} = \bar{b} = \bar{c} = \bar{d} = 0$, and therefore $a \equiv b \equiv c \equiv d \equiv 0 \pmod{k}$; hence $k^2 \mid kp$. But this last assertion implies $k \mid p$, which is impossible since $1 < k < p$; hence $l \neq 0$. By (11–2–1), (11–2–4), and (11–2–5), we see that

$$\begin{aligned}(kp)(kl) &= (a^2 + b^2 + c^2 + d^2)(\bar{a}^2 + \bar{b}^2 + \bar{c}^2 + \bar{d}^2)\\ &= (a\bar{a} + b\bar{b} + c\bar{c} + d\bar{d})^2 + (a\bar{b} - b\bar{a} - c\bar{d} + d\bar{c})^2\\ &\quad + (a\bar{c} + b\bar{d} - c\bar{a} - d\bar{b})^2 + (a\bar{d} - b\bar{c} + c\bar{b} - d\bar{a})^2.\end{aligned} \tag{11-2-6}$$

Each of the four expressions on the right hand side of (11–2–6) is a multiple of k; this is clear for the latter three, since $a \equiv \bar{a}$, $b \equiv \bar{b}$, $c \equiv \bar{c}$, and $d \equiv \bar{d} \pmod{k}$. For the first expression, we note that

$$a\bar{a} + b\bar{b} + c\bar{c} + d\bar{d} \equiv a^2 + b^2 + c^2 + d^2 = kp \equiv 0 \pmod{k}.$$

Consequently, there exist integers α, β, γ, and δ such that (11–2–6) may be rewritten as

$$k^2pl = (\alpha k)^2 + (\beta k)^2 + (\gamma k)^2 + (\delta k)^2,$$

or

$$lp = \alpha^2 + \beta^2 + \gamma^2 + \delta^2 \tag{11-2-7}$$

where $l < k$.

If k is even, we see that either all of a, b, c, and d are even, or two are even and two are odd, or all are odd. In any event, we can choose a, b, c, and d so that

$$a \equiv b \pmod{2},\ c \equiv d \pmod{2}.$$

Hence,

$$\frac{k}{2}p = \left(\frac{a-b}{2}\right)^2 + \left(\frac{a+b}{2}\right)^2 + \left(\frac{c-d}{2}\right)^2 + \left(\frac{c+d}{2}\right)^2.$$

Therefore, whether k is even or odd, if $k > 1$, we can find an integer less than k, say $\bar{k}$, such that $\bar{k}p$ is the sum of four nonnegative integral squares ($\bar{k} = l$ if k is odd, $\bar{k} = \frac{1}{2}k$ if k is even). However, since we chose k to be the least positive integer for which kp is the sum of four nonnegative integral squares, we have a contradiction. Therefore k must equal 1. Thus (11–2–4) holds with $k = 1$, and our proof is complete. ■

In this chapter we have discussed the circumstances under which the equations

$$x^2 + y^2 = n$$

and

$$x^2 + y^2 + z^2 + w^2 = n$$

have integral solutions. Theorems 11–1 and 11–3 form a small part of the theory of Diophantine equations.

After treating squares, we may inquire about the solvability of Diophantine equations involving cubes or higher powers. In 1770, E.

Waring stated that every integer is a sum of at most 9 (positive integral) cubes, a sum of at most 19 biquadrates, and so forth. This assertion has been interpreted to mean that for each integer $m \geq 2$, there is a corresponding integer N_m such that each positive integer n is a sum of at most N_m positive integral mth powers. D. Hilbert, one of the greatest mathematicians of the twentieth century, proved this assertion to be correct in 1909. His first proof depended on the use of a 25-fold integral. Succeeding mathematicians such as G. H. Hardy, J. E. Littlewood, and I. M. Vinogradov have made significant contributions to Waring's problem by determining sharp estimates for the size of N_m.

Perhaps the most well-known Diophantine equation is due to P. Fermat:

$$x^n + y^n = z^n. \tag{11-2-8}$$

Fermat noted in the margin of his copy of Diophantus' works:

> It is impossible to separate a cube into two cubes, or a biquadrate into two biquadrates, or in general any power higher than the second into powers of like degree; I have discovered a truly remarkable proof which this margin is too small to contain.

However, no proof that (11–2–8) lacks solutions when $n > 2$ has ever been found among Fermat's papers, nor has anyone been able to produce a proof. It has been shown that for $2 < n \leq 25{,}000$, (11–2–8) has no solutions; however, this is a long way from a complete proof.

EXERCISES

1. Prove (without assuming Corollary 10–1) that, if p is a prime $\equiv 1 \pmod 4$, then there exist positive integers m, x, and y such that

$$x^2 + y^2 = mp,$$

with $p \nmid x$, $p \nmid y$, $0 < m < p$. [Hint: Use the proof of Theorem 11–2.]

2. Let m_0 be the least possible m in Exercise 1. Prove that if $m_0 p = x_0^2 + y_0^2$, then m_0 does not divide both x_0 and y_0 unless $m_0 = 1$.

(In Exercises 3 through 7, assume $m_0 > 1$.)

3. Prove that there exist integers a and b such that, in the

notation of Exercise 2, $|x_0 - am_0| \leq \frac{1}{2} m_0$, $|y_0 - bm_0| \leq \frac{1}{2} m_0$, and $(x_0 - am_0)^2 + (y_0 - bm_0)^2 > 0$.

4. In the notation of Exercise 3, let $x_1 = x_0 - am_0$ and $y_1 = y_0 - bm_0$. Prove that $0 < x_1^2 + y_1^2 < m_0^2$, and $x_1^2 + y_1^2 = m_1 m_0$, where m_1 is some integer such that $0 < m_1 < m_0$.

5. In the notation of Exercise 4, prove that

$$\begin{aligned} m_0^2 m_1 p &= (x_0^2 + y_0^2)(x_1^2 + y_1^2) \\ &= (x_0 x_1 + y_0 y_1)^2 + (x_0 y_1 - x_1 y_0)^2. \end{aligned}$$

6. Prove that $m_0 \mid (x_0 x_1 + y_0 y_1)$ and $m_0 \mid (x_0 y_1 - x_1 y_0)$.

7. Letting $x_2 = (x_0 x_1 + y_0 y_1)/m_0$ and $y_2 = (x_0 y_1 - x_1 y_0)/m_0$, use Exercises 5, 6, and 7 to prove that $m_1 p = x_2^2 + y_2^2$.

8. Use the results of Exercises 1 through 7 to give a new proof that every prime $p \equiv 1 \pmod 4$ is a sum of two squares.

9. Prove that no integer of the form $4^a(8m + 7)$ is the sum of three squares. [Hint: Examine the congruence

$$x^2 + y^2 + z^2 \equiv 7 \pmod 8.]$$

CHAPTER 12

ELEMENTARY PARTITION THEORY

12–1 INTRODUCTION

The theory of partitions is an area of additive number theory, a subject concerning the representation of integers as sums of other integers. An elementary example of an additive theorem is the basis representation theorem (Theorem 1–3).

DEFINITION 12–1: A **partition** *of a nonnegative integer n is a representation of n as a sum of positive integers, called summands or parts of the partition. The order of the summands is irrelevant.*

Example 12–1: The partitions of 5 are 5, $4+1$, $3+2$, $3+1+1$, $2+2+1$, $2+1+1+1$, and $1+1+1+1+1$.

A sum such as $2+1+2$ is considered identical with $2+2+1$, since order is irrelevant. Thus, there are seven partitions of 5. Since order is irrelevant, we shall henceforth write partitions with nonincreasing order of summands.

We observe that 0 has one partition, the empty partition, and that the empty partition has no parts.

DEFINITION 12–2: *The function $p(n)$ will denote the number of partitions of n.*

Example 12–2: $p(0)=1$ and $p(5)=7$.

L. Euler was the first mathematician to discover important properties of $p(n)$; in 1748, he presented his results in his book *Introductio in Analysin Infinitorum.*

You may believe that the techniques of calculus have little relation to something defined as simply as $p(n)$. If so, a formula found by G. H. Hardy and S. Ramanujan in 1917 may surprise you. They obtained an exact expression for $p(n)$, the first term of which is

$$\frac{1}{2\pi\sqrt{2}}\,\frac{d}{dn}\left\{\frac{\exp\left(\dfrac{2\pi}{\sqrt{6}}\sqrt{n-\dfrac{1}{24}}\right)}{\sqrt{n-\dfrac{1}{24}}}\right\}. \tag{12-1-1}$$

As you may easily discern, $p(n)$ grows astronomically. Actual enumeration of the 3,972,999,029,388 partitions of 200 would certainly take more than a lifetime. However, the first five terms of the remarkable Hardy-Ramanujan formula give the correct value of $p(200)$.

To emphasize that $p(n)$ is not as simple as it appears, we note Ramanujan's congruence

$$p(5n+4) \equiv 0 \pmod 5. \tag{12-1-2}$$

The study of congruences such as (12–1–2) has involved some very deep properties of elliptic modular functions.

Partition identities, a subject we shall treat in great detail, is best exemplified by a theorem of Euler:

Theorem 12–1: *The number of partitions of an integer n in which all parts are odd equals the number of partitions of n in which all parts are distinct.*

Example 12–3: There are three partitions of 5 into odd parts: $5,\ 3+1+1,\ 1+1+1+1+1$; and three partitions of 5 into distinct parts: $5,\ 4+1,\ 3+2$.

Later in the chapter, we shall prove Euler's theorem and undertake discovery of new partition identities. First we consider a simple geometric device useful in partition theory.

12–2 GRAPHICAL REPRESENTATION

Some partition problems are solvable by graphical representation. In a graphical representation, a partition is represented by horizontal rows of dots. The graphical representations of the partitions of 5 are:

5	$4+1$	$3+2$	$3+1+1$
· · · · ·	· · · ·	· · ·	· · ·
	·	· ·	·
			·

$2+2+1$	$2+1+1+1$	$1+1+1+1+1$
. .	. .	.
. .	.	.
.	.	.
	.	.
		.

Note that the left-hand dots of each row lie on a vertical line and that the dots are equidistant.

Graphical representation allows us to introduce an important transformation of partitions.

DEFINITION 12–3: *The* **conjugate partition** *of a given partition is formed by reading the number of dots in the successive columns of the graphical representation.*

Example 12–4: The graphical representation for $5+4+1$, a partition of 10, is

```
. . . . .
. . . .
.
```

The conjugate partition, then, is

$$10 = \underbrace{3}_{\text{COL. 1}} + \underbrace{2}_{\text{COL. 2}} + \underbrace{2}_{\text{COL. 3}} + \underbrace{2}_{\text{COL. 4}} + \underbrace{1}_{\text{COL. 5}}$$

Example 12–5: The graphical representation for $8+4+3+1+1$, a partition of 17, is

```
. . . . . . . .
. . . .
. . .
.
.
```

The conjugate partition, then, is

$$17 = \underbrace{5}_{\text{COL. 1}} + \underbrace{3}_{\text{COL. 2}} + \underbrace{3}_{\text{COL. 3}} + \underbrace{2}_{\text{COL. 4}} + \underbrace{1}_{\text{COL. 5}} + \underbrace{1}_{\text{COL. 6}} + \underbrace{1}_{\text{COL. 7}} + \underbrace{1}_{\text{COL. 8}}$$

To illustrate the graphical technique, we shall prove the following theorem.

THEOREM 12–2: *If $p_m(n)$ denotes the number of partitions of* *n in which at most m parts appear, and if $\pi_m(n)$ denotes the number of partitions of n in which each part is no larger than m, then*

$$p_m(n) = \pi_m(n).$$

PROOF: Let us consider any partition of n in which at most m parts appear. The conjugate of such a partition has no parts larger than m since there could be at most m dots in any column of the graphical representation of the original partition.

This pairing of each partition of n of at most m parts with its conjugate, a partition of n in which no parts are larger than m, establishes a one-to-one correspondence between the two types of partitions. Hence, there must be the same number of each type; that is,

$$p_m(n) = \pi_m(n). \qquad \blacksquare$$

Example 12–6: Let $n = 5$ and $m = 3$; the five partitions of 5 with at most 3 parts are 5, 4 + 1, 3 + 2, 3 + 1 + 1, and 2 + 2 + 1, and the five partitions of 5 with no part exceeding 3 are 3 + 2, 3 + 1 + 1, 2 + 2 + 1, 2 + 1 + 1 + 1, and 1 + 1 + 1 + 1 + 1. The pairings are:

```
    5       <---->   1+1+1+1+1
  . . . . .              .
                         .
                         .
                         .
                         .

  4+1       <---->   2+1+1+1
  . . . .              . .
  .                    .
                       .
                       .

  3+2       <---->   2+2+1
  . . .                . .
  . .                  . .
                       .

  3+1+1     <---->   3+1+1
  . . .                . . .
  .                    .
  .                    .

  2+2+1     <---->   3+2
  . .                  . . .
  . .                  . .
  .
```

EXERCISES

1. Form the graphical representation of each of the following partitions, and find the conjugate partition in each case.

   (a) $6+4+3+2+1+1$ (d) $9+8+3+1$

   (b) $7+5+1$ (e) $8+6+2+2+1$

   (c) $10+7+6+3+3+3$ (f) $4+3+2+1$

2. List, for the case $n=6$, $m=4$, the pairings of partitions described in the proof of Theorem 12–2.

3. List, for the case $n=7$, $m=3$, the pairings of partitions described in the proof of Theorem 12–2.

4. A partition is called *self-conjugate* if it is identical with its conjugate. Thus, the two self-conjugate partitions of 8 are $4+2+1+1$ and $3+3+2$. The graphical representations of these partitions are

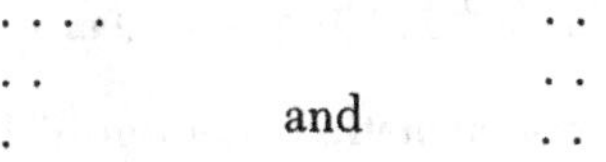

   respectively. We may now form two new partitions of 8 by uniting dots lying on successive right angles as follows:

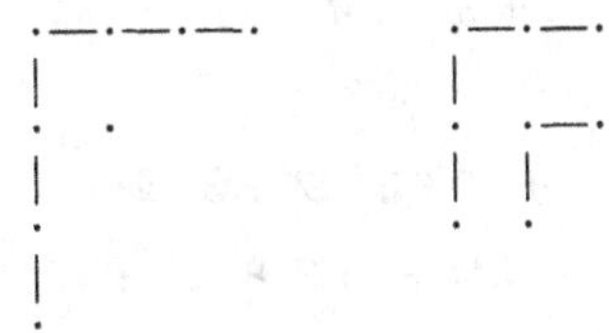

   This produces $7+1$ and $5+3$. Prove that this procedure establishes that the number of self-conjugate partitions of n always equals the number of partitions of n into distinct odd parts.

5. Let $\mathcal{O}(n)$ denote the number of partitions of n with distinct odd parts. Utilize Exercise 4 to prove that $p(n)-\mathcal{O}(n)$ is the number of partitions of n that are not self-conjugate.

6. Use Exercise 5 to prove that $p(n) \equiv \mathcal{O}(n) \pmod 2$.

12-3 EULER'S PARTITION THEOREM

We shall now establish the theorem of Euler stated in Section 12-1.

Theorem 12-3: *The number of partitions of an integer n in which all parts are odd equals the number of partitions of n in which all parts are distinct.*

Proof: As in Theorem 12-2, we shall establish a one-to-one correspondence between the two types of partitions.

Let us first consider a partition having only odd parts. If f_i denotes the number of times i appears as a part, we may write our partition as

$$n = \underbrace{1+1+\ldots+1}_{f_1 \text{ times}} + \underbrace{3+3+\ldots+3}_{f_3 \text{ times}} + \underbrace{5+5+\ldots+5}_{f_5 \text{ times}} + \ldots + \underbrace{(2M-1)+(2M-1)+\ldots+(2M-1)}_{f_{2M-1} \text{ times}}$$

$$= f_1 \cdot 1 + f_3 \cdot 3 + f_5 \cdot 5 + \ldots + f_{2M-1} \cdot (2M-1). \qquad \textbf{(12-3-1)}$$

By the basis representation theorem (Theorem 1–3), we know that each f_i can be represented uniquely as a sum of distinct powers of 2. Thus,

$$n = (2^a + 2^b + \ldots + 2^c)1 + (2^e + 2^f + \ldots + 2^g)3 + \ldots + (2^r + 2^s + \ldots + 2^t)(2M-1), \qquad \textbf{(12-3-2)}$$

and so

$$n = 2^a + 2^b + \ldots + 2^c + 3 \cdot 2^e + 3 \cdot 2^f + \ldots + 3 \cdot 2^g + \ldots + (2M-1)2^r + (2M-1)2^s + \ldots + (2M-1)2^t. \qquad \textbf{(12-3-3)}$$

This last expression is a partition of n into distinct parts, for we know by the fundamental theorem of arithmetic that numbers with different powers of 2 in their prime factorization, or with distinct largest odd factors, are distinct numbers.

We must now verify that we have established a one-to-one correspondence between the partitions of n with odd parts and those with distinct parts. Suppose we start with a partition of n into distinct parts. Then we can write each part as the product of an odd number and a power of 2; we thus obtain an expression like (12-3-3). We now collect all parts with identical largest odd factors. Then, factoring out the various odd factors, we obtain (12–3–2). By merely summing the various powers of 2 in (12–3–2), we obtain a partition of n with only odd parts as in (12–3–1). Thus, having established our one-to-one correspondence, we have proved the theorem. ■

Example 12–7: When $n = 5$, the pairing we have established in the proof of Theorem 12–3 is

$$5 \longleftrightarrow 5$$

$$3+1+1=2\cdot 1+3 \longleftrightarrow 3+2$$

$$1+1+1+1+1=5\cdot 1 \longleftrightarrow (4+1)\cdot 1=4+1.$$

EXERCISES

1. For the case $n=9$, list the pairings of partitions described in the proof of Theorem 12–3.

2. Prove that the number of partitions of n in which no part is repeated more than $k-1$ times equals the number of partitions of n in which no part is divisible by k. [Hint: Follow the proof of Theorem 12–3, using representation of f_i to the base k rather than to the base 2.]

12–4 SEARCHING FOR PARTITION IDENTITIES

In this section we shall attempt to ferret out results similar to Euler's Partition Theorem. Our object here is not *proof*, but *discovery*.

To illustrate how we conduct the search, we forget Theorem 12–3 and attempt to discover it.

DEFINITION 12–4: *If S is a set of positive integers, let $p(S,n)$ denote the number of partitions of n in which each summand is an element of S.*

Problem 12–1 (Discovering Euler's Partition Theorem): If $D_1(n)$ denotes the number of partitions of n into distinct parts, can we find a set S_1 such that

$$p(S_1, n) = D_1(n)$$

for all n?

Since $D_1(1) = 1$, we see that $1 \in S_1$ (otherwise $p(S_1,1) = 0$). Since $D_1(2) = 1$, we see that $2 \notin S_1$ (otherwise $p(S_1,2) = 2$). Since $D_1(3) =$

TABLE 12–1: DISCOVERING ELEMENTS OF S_1.

n	$D_1(n)$	$p(S_1,n)$ if $n \notin S_1$	$p(S_1,n)$ if $n \in S_1$	Conclusion
1	1	0	1	$1 \in S_1$
2	1	1	2	$2 \notin S_1$
3	2	1	2	$3 \in S_1$
4	2	2	3	$4 \notin S_1$
5	3	2	3	$5 \in S_1$
6	4	4	5	$6 \notin S_1$
7	5	4	5	$7 \in S_1$

2, we see that $3 \in S_1$ (otherwise $p(S_1,3) = 1$). Since $D_1(4) = 2$, we see that $4 \notin S_1$ (otherwise $p(S_1,4) = 3$). Since $D_1(5) = 3$, we see that $5 \in S_1$ (otherwise $p(S_1,5) = 2$). Since $D_1(6) = 4$, we see that $6 \notin S_1$ (otherwise $p(S_1,6) = 5$). Since $D_1(7) = 5$, we see that $7 \in S_1$ (otherwise $p(S_1,7) = 4$). Table 12–1 contains a summary of these observations. It is thus a reasonable guess that S_1 is the set of all odd numbers.

We shall now use the same approach to "discover" a famous theorem of L. J. Rogers (also discovered independently by S. Ramanujan and I. J. Schur).

Problem 12–2 (Discovering the First Rogers-Ramanujan Identity): Let $D_2(n)$ denote the number of partitions of n in which any two summands differ by at least 2. Can we find a set of integers S_2 such that

$$p(S_2,n) = D_2(n)?$$

Let us form a table, proceeding as before.

TABLE 12–2: DISCOVERING ELEMENTS OF S_2.

n	$D_2(n)$	$p(S_2,n)$ if $n \in S_2$	$p(S_2,n)$ if $n \notin S_2$	Conclusion
1	1	1	0	$1 \in S_2$
2	1	2	1	$2 \notin S_2$
3	1	2	1	$3 \notin S_2$
4	2	2	1	$4 \in S_2$
5	2	3	2	$5 \notin S_2$
6	3	3	2	$6 \in S_2$
7	3	4	3	$7 \notin S_2$
8	4	5	4	$8 \notin S_2$
9	5	5	4	$9 \in S_2$
10	6	7	6	$10 \notin S_2$
11	7	7	6	$11 \in S_2$
12	9	10	9	$12 \notin S_2$
13	10	11	10	$13 \notin S_2$
14	12	12	11	$14 \in S_2$
15	14	15	14	$15 \notin S_2$
16	17	17	16	$16 \in S_2$

We see that 1, 4, 6, 9, 11, 14, 16 are the first seven elements of S_2. So far, these are all the numbers one unit away from a multiple of 5. Thus, we may conjecture that S_2 consists of all positive integers congruent to either 1 or 4 modulo 5. We shall see in Chapter 14 that this is indeed correct.

DEFINITION 12-5: *$D_d(n)$ will denote the number of partitions of n in which any two summands differ by at least d.*

We note that the previously defined quantities $D_1(n)$ and $D_2(n)$ represent the special cases of $D_d(n)$ for $d=1$ and $d=2$.

To illustrate some of the pitfalls of searching for partition identities, we proceed to study $D_3(n)$.

Problem 12-3: Can we find a set of integers $\mathcal{S}_3$ such that

$$p(\mathcal{S}_3,n) = D_3(n)?$$

Again we construct a table.

TABLE 12-3: DISCOVERING ELEMENTS OF $\mathcal{S}_3$.

n	$D_3(n)$	$p(\mathcal{S}_3,n)$ if $n \in \mathcal{S}_3$	$p(\mathcal{S}_3,n)$ if $n \notin \mathcal{S}_3$	Conclusion
1	1	1	0	$1 \in \mathcal{S}_3$
2	1	2	1	$2 \notin \mathcal{S}_3$
3	1	2	1	$3 \notin \mathcal{S}_3$
4	1	2	1	$4 \notin \mathcal{S}_3$
5	2	2	1	$5 \in \mathcal{S}_3$
6	2	3	2	$6 \notin \mathcal{S}_3$
7	3	3	2	$7 \in \mathcal{S}_3$
8	3	4	3	$8 \notin \mathcal{S}_3$
9	4	4	3	$9 \in \mathcal{S}_3$
10	4	6	5	?

Thus we see that $\mathcal{S}_3$ cannot exist, a fact that may be formulated in the following theorem.

THEOREM 12-4: *There exists no set $\mathcal{S}_3$ of integers such that*

$$p(\mathcal{S}_3,n) = D_3(n).$$

It may seem that we are stymied in finding a new positive result; however, there is a result in which a partition function only slightly different from $D_3(n)$ is considered.

If S_3 denotes the set of integers congruent to 1 or 5 modulo 6, and if $\bar{D}_3(n)$ denotes the number of partitions of n in which the difference between any two parts is at least 3, and in which no consecutive multiples of 3 appear, then $p(S_3,n) = \bar{D}_3(n)$.

This result, originally discovered by I. J. Schur, will be proved in Chapter 14. Schur's result is the first indication that, in order to obtain valid theorems, we need to introduce new partition functions like $\bar{D}_3(n)$.

To prove Schur's theorem and the theorem of Rogers and Ramanujan, we must first study generating functions.

EXERCISES *(In the* jth *exercise below, attempt to define a set of integers* W_j *such that* $E_j(n) = p(W_j,n)$ *for all* n.*)*

1. $E_1(n)$ denotes the number of partitions of n in which all parts differ by at least 2 and no consecutive even numbers appear as summands.

2. $E_2(n)$ denotes the number of partitions of n in which all parts differ by at least 2 and no consecutive odd numbers appear as summands.

3. $E_3(n)$ denotes the number of partitions of n in which all parts differ by at least 2, no part is less than 3, and no consecutive even numbers appear as summands.

4. $E_4(n)$ denotes the number of partitions of n in which all parts differ by at least 2, no part is less than 2, and no consecutive odd numbers appear as summands.

5. $E_5(n)$ denotes the number of partitions of n in which no part appears more than twice.

6. $E_6(n)$ denotes the number of partitions of n in which the difference between any odd part and any other part not exceeding it (if such exists) is at least 3, and no part is less than 2.

7. $E_7(n)$ denotes the number of partitions of n in which no part appears more than twice; and, if a part m appears twice, neither $m-1$ nor $m+1$ appears at all.

8. $E_8(n)$ denotes the number of partitions of n in which all parts differ by at least 6, neither 1 nor 3 appears as a part, and, if m and $m+6$ appear as parts, then $m \not\equiv 0, 1, 3 \pmod 6$.

9. $E_9(n)$ denotes the number of partitions of n in which the difference between any odd part and any other part not exceeding it (if such exists) is at least 1.

10. $E_{10}(n)$ denotes the number of partitions of n in which no part appears more than thrice with the restriction that, if any part m appears 2 or 3 times, then $m+1$ appears at most once.

11. $E_{11}(n)$ denotes the number of partitions of n in which no part is less than 2, no part appears more than twice, and no consecutive integers appear as summands.

12. $E_{12}(n)$ denotes the number of partitions of n into parts such that no consecutive integers appear as summands, and no part is less than 2.

13. $E_{13}(n)$ denotes the number of partitions of n in which no part appears more than once and no multiple of 3 appears.

CHAPTER 13

PARTITION GENERATING FUNCTIONS

In Chapter 12, we used purely elementary procedures to establish theorems about various partition functions. We can obtain far more interesting results about partitions by using generating functions, many of which are infinite products. An infinite product is defined in much the same way as an infinite series.

DEFINITION 13–1: *If $a_0, a_1, a_2, a_3, \ldots$ is a sequence of numbers, then the nth* **partial product** *of the a_i is*

$$\prod_{j=0}^{n} a_j = a_0 a_1 a_2 \ldots a_n,$$

and the **infinite product** *of the a_i is*

$$\prod_{j=0}^{\infty} a_j = \lim_{n\to\infty} \prod_{j=0}^{n} a_j,$$

if that limit exists and is not zero. Otherwise, the product is said to diverge.

We shall first study the relationship between infinite products and partition functions. Afterwards, we shall examine some theorems relating infinite products and infinite series; these results will be valuable to our understanding of partition identities in Chapter 14.

13–1 INFINITE PRODUCTS AS GENERATING FUNCTIONS

THE GENERATING FUNCTION FOR $\pi_m(n)$

Theorem 13–1 illustrates a relationship between products and partition functions by exhibiting the generating function for $\pi_m(n)$, the number of partitions of n in which no part is greater than m.

THEOREM 13-1: $\sum_{n=0}^{\infty} \pi_m(n) q^n = \prod_{j=1}^{m} \frac{1}{1-q^j}$, *where* $|q| < 1$.

PROOF: Recall the formula for the sum of a geometric series,

$$\frac{1}{1-x} = 1 + x + x^2 + x^3 + x^4 + \ldots, \text{ where } |x| < 1.$$

(We can obtain this by letting $n \to \infty$ in Theorem 1-2.)
Thus,

$$\begin{aligned}
\prod_{j=1}^{m} \frac{1}{1-q^j} &= \frac{1}{1-q}\,\frac{1}{1-q^2}\,\frac{1}{1-q^3}\cdots\frac{1}{1-q^m} \\
&= (1 + q^1 + q^{2\cdot 1} + q^{3\cdot 1} + q^{4\cdot 1} + \ldots) \\
&\quad \cdot (1 + q^2 + q^{2\cdot 2} + q^{3\cdot 2} + q^{4\cdot 2} + \ldots) \\
&\quad \cdot (1 + q^3 + q^{2\cdot 3} + q^{3\cdot 3} + q^{4\cdot 3} + \ldots) \\
&\quad \cdots \\
&\quad \cdot (1 + q^m + q^{2\cdot m} + q^{3\cdot m} + q^{4\cdot m} + \ldots).
\end{aligned}$$

Now, since the number of these series is finite and all converge absolutely for $|q| < 1$, we may multiply them and collect terms. Hence,

$$\begin{aligned}
\prod_{j=1}^{m} \frac{1}{1-q^j} &= 1 + q^1 + (q^{2\cdot 1} + q^2) \\
&\quad + (q^{3\cdot 1} + q^{2+1} + q^3) \\
&\quad + (q^{4\cdot 1} + q^{2+2\cdot 1} + q^{2\cdot 2} + q^{3+1} + q^4) \\
&\quad + \ldots.
\end{aligned}$$

Let us examine one of the lines above, say $q^{4\cdot 1} + q^{2+2\cdot 1} + q^{2\cdot 2} + q^{3+1} + q^4$. The exponents are $4\cdot 1$ $(=1+1+1+1)$, $2+2\cdot 1$ $(=2+1+1)$, $2\cdot 2$ $(=2+2)$, $3+1$, and 4; these are precisely the partitions of 4. In the general case, the expression for each exponent of sum n contains $\pi_m(n)$ terms, one corresponding to each way of writing the exponent n as a sum of the form $f_1 \cdot 1 + f_2 \cdot 2 + \ldots + f_m \cdot m$. This last expression is merely the partition of n into parts each not exceeding m, where 1 appears f_1 times, 2 appears f_2 times, . . . , and m appears f_m times. Thus,

$$\prod_{j=1}^{m} \frac{1}{1-q^j} = \sum_{n=0}^{\infty} \pi_m(n) q^n. \qquad \blacksquare$$

THE GENERATING FUNCTION FOR $p(n)$

To study the generating function $\sum_{n=0}^{\infty} p(n)q^n$ for $p(n)$, we need the following result.

THEOREM 13–2: $\lim_{n\to\infty} p(n)^{1/n} = 1.$

This theorem and the root test of calculus imply that $\sum_{n=0}^{\infty} p(n)q^n$ converges for $|q| < 1$. Rather than disrupt our discussion, we refer interested readers to Appendix A for a complete proof.

THEOREM 13–3: $\sum_{n=0}^{\infty} p(n)q^n = \prod_{j=1}^{\infty} \frac{1}{1-q^j}$, *where* $|q| < 1$.

PROOF: Note that if $m \geq n$, then $\pi_m(n) = p(n)$, since no part of a partition can exceed the number being partitioned. Clearly, $0 \leq \pi_m(n) \leq p(n)$ for all m and n. Consequently,

$$\left| \sum_{n=0}^{\infty} p(n)q^n - \prod_{j=1}^{m} \frac{1}{1-q^j} \right| = \left| \sum_{n=0}^{\infty} p(n)q^n - \sum_{n=0}^{\infty} \pi_m(n)q^n \right|$$

$$= \left| \sum_{n=m+1}^{\infty} (p(n) - \pi_m(n))q^n \right|$$

$$\leq \sum_{n=m+1}^{\infty} p(n)\ |q|^n.$$

Since Theorem 13–2 shows that the series $\sum_{n=0}^{\infty} p(n)q^n$ converges for $|q| < 1$, we see that

$$\lim_{m\to\infty} \sum_{n=m+1}^{\infty} p(n)|q|^n = 0;$$

hence,

$$\sum_{n=0}^{\infty} p(n)q^n = \lim_{m\to\infty} \prod_{j=1}^{m} \frac{1}{1-q^j} = \prod_{j=1}^{\infty} \frac{1}{1-q^j}. \qquad \blacksquare$$

THE GENERATING FUNCTIONS FOR $p(S_d,n)$

In exactly the same way, one may establish the formulae:

$$\sum_{n=0}^{\infty} p(S_1,n)\,q^n = \prod_{j=1}^{\infty} \frac{1}{1-q^{2j-1}}, \tag{13-1-1}$$

$$\sum_{n=0}^{\infty} p(S_2,n)\,q^n = \prod_{j=1}^{\infty} \frac{1}{(1-q^{5j-4})(1-q^{5j-1})}, \tag{13-1-2}$$

$$\sum_{n=0}^{\infty} p(S_3,n)\,q^n = \prod_{j=1}^{\infty} \frac{1}{(1-q^{6j-5})(1-q^{6j-1})}, \tag{13-1-3}$$

where S_d is the set of all positive integers congruent to 1 or to $d+2$ modulo $d+3$; the series and products converge for $|q| < 1$.

THE GENERATING FUNCTION FOR $d_m(n)$

By slightly altering the proofs of Theorems 13–1 and 13–3, we can find the generating function for partitions with distinct parts.

THEOREM 13–4: *If $d_m(n)$ denotes the number of partitions of n into distinct parts, none greater than m, then*

$$\sum_{n=0}^{\infty} d_m(n)\,q^n = \prod_{j=1}^{m} (1+q^j).$$

PROOF: $$\prod_{j=1}^{m} (1+q^j) = (1+q)(1+q^2)(1+q^3)\ldots(1+q^m)$$

$$= 1 + q + q^2 + (q^{2+1} + q^3) + (q^4 + q^{3+1})$$

$$+ (q^5 + q^{4+1} + q^{3+2}) + \ldots.$$

In general, the term q^n appears as often as n can be expressed as a sum of distinct integers, each not exceeding m. Hence,

$$\prod_{j=1}^{m} (1+q^j) = \sum_{n=0}^{\infty} d_m(n)\,q^n. \qquad \blacksquare$$

THE GENERATING FUNCTION FOR $D_1(n)$

We can now obtain the generating function for $D_1(n)$ from Theorem 13–4, just as we obtained the generating function for $p(n)$ from Theorem 13–1.

THEOREM 13–5: $\sum_{n=0}^{\infty} D_1(n)q^n = \prod_{j=1}^{\infty}(1+q^j)$, *where* $|q|<1$.

PROOF: As in the proof of Theorem 13–3, we note that $d_m(n) = D_1(n)$, if $m \geq n$. Also, $0 \leq d_m(n) \leq D_1(n) \leq p(n)$ for all m and n. Consequently,

$$\begin{aligned}\left|\sum_{n=0}^{\infty} D_1(n)q^n - \prod_{j=1}^{m}(1+q^j)\right| &= \left|\sum_{n=0}^{\infty} D_1(n)q^n - \sum_{n=0}^{\infty} d_m(n)q^n\right| \\ &= \left|\sum_{n=m+1}^{\infty}(D_1(n)-d_m(n))q^n\right| \\ &\leq \sum_{n=m+1}^{\infty} D_1(n)|q|^n \\ &\leq \sum_{n=m+1}^{\infty} p(n)|q|^n.\end{aligned}$$

As in the proof of Theorem 13–1, we see that

$$\lim_{m\to\infty}\sum_{n=m+1}^{\infty} p(n)|q|^n = 0$$

for $|q|<1$; hence

$$\sum_{n=0}^{\infty} D_1(n)q^n = \lim_{m\to\infty}\prod_{j=1}^{m}(1+q^j) = \prod_{j=1}^{\infty}(1+q^j). \qquad \blacksquare$$

SOME APPLICATIONS OF GENERATING FUNCTIONS

We shall now see the utility of generating functions in establishing relations between various partition functions. First we give a different proof of Theorem 12–3.

$$\begin{aligned}\sum_{n=0}^{\infty} D_1(n)q^n &= \prod_{j=1}^{\infty}(1+q^j) \\ &= \prod_{j=1}^{\infty}\frac{(1+q^j)(1-q^j)}{(1-q^j)} \\ &= \prod_{j=1}^{\infty}\frac{(1-q^{2j})}{(1-q^j)}\end{aligned}$$

$$= \prod_{j=1}^{\infty} (1 - q^{2j}) \prod_{j=1}^{\infty} \frac{1}{(1 - q^{j})}$$

$$= \prod_{j=1}^{\infty} (1 - q^{2j}) \prod_{j=1}^{\infty} \frac{1}{(1 - q^{2j})(1 - q^{2j-1})}$$

$$= \prod_{j=1}^{\infty} \frac{1}{(1 - q^{2j-1})}$$

$$= \sum_{n=0}^{\infty} p(S_1, n) q^n.$$

Now, as we remarked in Section 3–4, a function has at most one MacLaurin series expansion. Consequently, $D_1(n) = p(S_1, n)$ for all n. ■

This proof of Euler's Partition Theorem relies on manipulations and rearrangements of infinite products. The correctness of such procedures is established in Appendix B.

We can now prove a result of I. Schur that also relies on the manipulation of infinite products.

THEOREM 13–6: *If $Q_3(n)$ denotes the number of partitions of n into distinct parts, no part being a multiple of 3, then $Q_3(n) = p(S_3, n)$ for all n.*

PROOF: As with $D_1(n)$, we can prove that

$$\sum_{n=0}^{\infty} Q_3(n) q^n = \prod_{j=1}^{\infty} (1 + q^{3j-2})(1 + q^{3j-1}).$$

Hence,

$$\sum_{n=0}^{\infty} Q_3(n) q^n = \prod_{j=1}^{\infty} (1 + q^{3j-2})(1 + q^{3j-1})$$

$$= \prod_{j=1}^{\infty} \frac{(1 + q^{3j-1})(1 + q^{3j-2})(1 - q^{3j-1})(1 - q^{3j-2})}{(1 - q^{3j-1})(1 - q^{3j-2})}$$

$$= \prod_{j=1}^{\infty} \frac{(1 - q^{6j-2})(1 - q^{6j-4})}{(1 - q^{3j-1})(1 - q^{3j-2})}$$

$$= \prod_{j=1}^{\infty} (1 - q^{6j-2})(1 - q^{6j-4}) \prod_{j=1}^{\infty} \frac{1}{(1 - q^{3j-1})(1 - q^{3j-2})}$$

$$= \prod_{j=1}^{\infty} (1 - q^{6j-2})(1 - q^{6j-4})$$

$$\prod_{j=1}^{\infty} \frac{1}{(1 - q^{6j-4})(1 - q^{6j-1})(1 - q^{6j-5})(1 - q^{6j-2})}$$

$$= \prod_{j=1}^{\infty} \frac{1}{(1 - q^{6j-5})(1 - q^{6j-1})}$$

$$= \sum_{n=0}^{\infty} p(S_3, n) q^n;$$

thus $Q_3(n) = p(S_3, n)$. ∎

EXERCISES

1. Prove directly (that is, without using Theorem 13–1) that

$$\pi_m(n) = \pi_m(n - m) + \pi_{m-1}(n),$$

where $\pi_m(n)$ is defined to be 1 if $n = 0$ and 0 if $n < 0$.

2. Deduce from Exercise 1 that if $F_m(q) = \sum_{n=0}^{\infty} \pi_m(n) q^n$, then

$$F_m(q) = q^m F_m(q) + F_{m-1}(q).$$

3. Use Exercise 2 to give a new proof of Theorem 13–1.

4. Prove directly (that is, not using Theorem 13–4) that

$$d_m(n) = d_{m-1}(n) + d_{m-1}(n - m),$$

where $d_m(n)$ is defined to be 1 if $n = 0$ and 0 if $n < 0$.

5. Deduce from Exercise 4 that if $L_m(q) = \sum_{n=0}^{\infty} d_m(n) q^n$, then

$$L_m(q) = L_{m-1}(q) + q^m L_{m-1}(q).$$

6. Use Exercise 5 to give a new proof of Theorem 13–4.

7. Prove that the number of partitions of n into distinct parts congruent to 1, 2, or 4 (mod 7) equals the number of partitions of n into parts congruent to 1, 9, or 11 (mod 14).

8. Let S be any set of integers such that $2j \in S$ whenever $j \in S$. Define S' to be the set of integers $j \in S$ for which $\frac{j}{2} \notin S$. Prove that the number of partitions of an integer n into parts taken from S' equals the number of partitions of n into distinct parts taken from S.

13–2 IDENTITIES BETWEEN INFINITE SERIES AND PRODUCTS

Here we shall prove some basic results relating infinite series to infinite products. Throughout this section, we shall ignore questions of convergence and rearrangement of series; these receive full treatment in Appendix B.

Theorem 13–7: *If $|q| < 1$, then*

$$1+\sum_{n=1}^{\infty} \frac{q^{n(n-1)/2}z^n}{(1-q)(1-q^2)\ldots(1-q^n)} = \prod_{n=0}^{\infty} (1+zq^n) \tag{13-2-1}$$

and

$$1+\sum_{n=1}^{\infty} \frac{z^n}{(1-q)(1-q^2)\ldots(1-q^n)} = \prod_{n=0}^{\infty} \frac{1}{(1-zq^n)}, \tag{13-2-2}$$

with the stipulation that $|z| < 1$ in (13–2–2).

Proof: We start with (13–2–1). The infinite product in (13–2–1) must have a MacLaurin series expansion in z (see Appendix B for a proof). Let

$$f(z) = \prod_{n=0}^{\infty} (1+zq^n), \tag{13-2-3}$$

and let the MacLaurin series expansion for $f(z)$ be

$$f(z) = \sum_{n=0}^{\infty} A_n z^n, \tag{13-2-4}$$

where the A_n depend on q. Now

$$f(z) = \prod_{n=0}^{\infty} (1+zq^n)$$

$$= (1 + z) \prod_{n=1}^{\infty} (1 + zq^n)$$

$$= (1 + z) \prod_{n=0}^{\infty} (1 + zq^{n+1})$$

$$= (1 + z) \prod_{n=0}^{\infty} (1 + zqq^n)$$

$$= (1 + z)f(zq). \tag{13-2-5}$$

Substituting (13–2–4) into (13–2–5), we obtain the relations

$$\sum_{n=0}^{\infty} A_n z^n = (1 + z) \sum_{n=0}^{\infty} A_n z^n q^n$$

$$= \sum_{n=0}^{\infty} A_n z^n q^n + \sum_{n=0}^{\infty} A_n z^{n+1} q^n. \tag{13-2-6}$$

Now, since A_0 is the constant term of the MacLaurin series, $A_0 = f(0) = 1$. For $N > 0$, let us compare the coefficient of a general power, say z^N, on both sides of (13–2–6). On the left side it is A_N; on the right, it is $A_N q^N + A_{N-1} q^{N-1}$. Since a function, in this case $f(z)$, has at most one MacLaurin series expansion, we find that

$$A_N = A_N q^N + A_{N-1} q^{N-1}.$$

Therefore,

$$(1 - q^N) A_N = A_{N-1} q^{N-1};$$

that is,

$$A_N = \frac{q^{N-1}}{(1 - q^N)} A_{N-1}. \tag{13-2-7}$$

By repeated application of (13–2–7), we can express A_N in terms of q, as follows:

$$A_N = \frac{q^{N-1}}{(1 - q^N)} A_{N-1}$$

$$= \frac{q^{N-1}}{(1 - q^N)} \frac{q^{N-2}}{(1 - q^{N-1})} A_{N-2}$$

$$= \frac{q^{N-1}}{(1 - q^N)} \frac{q^{N-2}}{(1 - q^{N-1})} \frac{q^{N-3}}{(1 - q^{N-2})} A_{N-3}$$

$$= \frac{q^{(N-1)+(N-2)+(N-3)+\ldots+2+1+0}}{(1-q^N)(1-q^{N-1})\ldots(1-q)} A_0.$$

The sum of the first $N-1$ positive integers is $(N^2-N)/2$, by Theorem 1-1, and $A_0 = 1$. Thus we obtain the formula

$$A_N = \frac{q^{(N^2-N)/2}}{(1-q^N)(1-q^{N-1})\ldots(1-q)}.$$

Hence,

$$\begin{aligned}\prod_{n=0}^{\infty}(1+zq^n) &= f(z)\\ &= \sum_{n=0}^{\infty} A_n z^n\\ &= 1 + \sum_{n=1}^{\infty} \frac{q^{(n^2-n)/2} z^n}{(1-q)(1-q^2)\ldots(1-q^n)},\end{aligned}$$

as we asserted in (13–2–1).

Since the proof of (13–2–2) mirrors the proof of (13–2–1), we shall merely outline the procedure. Let

$$g(z) = \sum_{n=0}^{\infty} B_n z^n = \prod_{n=0}^{\infty} \frac{1}{(1-zq^n)}. \tag{13-2-8}$$

Then

$$(1-z)g(z) = g(zq). \tag{13-2-9}$$

Consequently, $B_0 = 1$ and

$$(1-q^n)B_n = B_{n-1}. \tag{13-2-10}$$

Therefore

$$B_n = \frac{1}{(1-q)(1-q^2)\ldots(1-q^n)} \text{ for } n \geq 1. \tag{13-2-11}$$

Substituting (13–2–11) into (13–2–8), we obtain (13–2–2). ■

Our next result is known as Jacobi's Triple Product Identity.

Theorem 13–8: *If $z \neq 0$ and $|q| < 1$, then*

$$\prod_{n=0}^{\infty} (1 - q^{2n+2})(1 + zq^{2n+1})(1 + z^{-1}q^{2n+1}) = \sum_{n=-\infty}^{\infty} q^{n^2}z^n. \quad \textbf{(13-2-12)}$$

REMARK: We can think of the doubly infinite series $\sum_{n=-\infty}^{\infty} q^{n^2}z^n$ as the sum

$$\sum_{n=0}^{\infty} q^{n^2}z^n + \sum_{n=-1}^{-\infty} q^{n^2}z^n$$

of two infinite series. The last expression may look more familiar if in the second series we replace n by $-n$. Thus,

$$\sum_{n=-\infty}^{\infty} q^{n^2}z^n = \sum_{n=0}^{\infty} q^{n^2}z^n + \sum_{n=1}^{\infty} q^{n^2}z^{-n}.$$

PROOF: Assume $|z| > |q|$. In (13–2–1), replace q by q^2 and z by zq. This yields the formula

$$\prod_{n=0}^{\infty} (1 + zq^{2n+1}) = 1 + \sum_{n=1}^{\infty} \frac{q^{n^2}z^n}{(1-q^2)(1-q^4)\ldots(1-q^{2n})}. \quad \textbf{(13-2-13)}$$

Now

$$\frac{1}{(1-q^2)(1-q^4)\ldots(1-q^{2n})} = \frac{\prod_{m=0}^{\infty} (1 - q^{2m+2n+2})}{\prod_{m=0}^{\infty} (1 - q^{2m+2})}.$$

Therefore,

$$\prod_{n=0}^{\infty} (1 + zq^{2n+1}) = \sum_{n=0}^{\infty} q^{n^2}z^n \frac{\prod_{m=0}^{\infty} (1 - q^{2m+2n+2})}{\prod_{m=0}^{\infty} (1 - q^{2m+2})}.$$

Since the product in the denominator does not involve n, we may factor it out of the sum and obtain the formula

$$\prod_{n=0}^{\infty} (1 + zq^{2n+1}) = \prod_{m=0}^{\infty} \frac{1}{(1-q^{2m+2})} \sum_{n=0}^{\infty} q^{n^2}z^n \prod_{m=0}^{\infty} (1 - q^{2m+2n+2}). \quad \textbf{(13-2-14)}$$

Note that if n is a negative integer, then

$$\prod_{m=0}^{\infty} (1 - q^{2m+2n+2}) = 0,$$

since, for this product, the term with $m = -n - 1$ yields

$$(1 - q^{-2n-2+2n+2}) = (1 - q^0) = 1 - 1 = 0.$$

Hence, if we extend the range of the sum in (13–2–14) from $0 \leq n < \infty$ to $-\infty < n < \infty$, we do not alter the value of the series, since we have introduced only terms whose values are zero. Thus,

$$\prod_{n=0}^{\infty} (1 + zq^{2n+1}) = \prod_{m=0}^{\infty} \frac{1}{(1 - q^{2m+2})} \sum_{n=-\infty}^{\infty} q^{n^2} z^n \prod_{m=0}^{\infty} (1 - q^{2m+2n+2}). \qquad (13\text{-}2\text{-}15)$$

Now, if in (13–2–1) we replace n by m and q by q^2, and then set $z = -q^{2n+2}$, we have the formula

$$\prod_{m=0}^{\infty} (1 - q^{2m+2n+2}) = 1 + \sum_{m=1}^{\infty} \frac{(-1)^m q^{m^2+2mn+m}}{(1 - q^2) \ldots (1 - q^{2m})}. \qquad (13\text{-}2\text{-}16)$$

Thus,

$$\prod_{n=0}^{\infty} (1 + zq^{2n+1}) = \prod_{m=0}^{\infty} \frac{1}{(1 - q^{2m+2})} \sum_{n=-\infty}^{\infty} q^{n^2} z^n \left(1 + \sum_{m=1}^{\infty} \frac{(-1)^m q^{m^2+2mn+m}}{(1 - q^2) \ldots (1 - q^{2m})}\right)$$

$$= \prod_{m=0}^{\infty} \frac{1}{(1 - q^{2m+2})} \left(\sum_{n=-\infty}^{\infty} q^{n^2} z^n + \sum_{m=1}^{\infty} \frac{(-1)^m q^m}{(1 - q^2) \ldots (1 - q^{2m})} \sum_{n=-\infty}^{\infty} q^{(n+m)^2} z^n\right)$$

$$= \prod_{m=0}^{\infty} \frac{1}{(1 - q^{2m+2})} \left(\sum_{n=-\infty}^{\infty} q^{n^2} z^n + \sum_{m=1}^{\infty} \frac{(-1)^m q^m z^{-m}}{(1 - q^2) \ldots (1 - q^{2m})} \sum_{n=-\infty}^{\infty} q^{(n+m)^2} z^{n+m}\right). \qquad (13\text{-}2\text{-}17)$$

Note that for each integer m,

$$\sum_{n=-\infty}^{\infty} q^{(n+m)^2} z^{n+m} = \sum_{n=-\infty}^{\infty} q^{n^2} z^n,$$

since both $n+m$ and n assume each integral value exactly once. Thus,

$$\prod_{n=0}^{\infty}(1+zq^{2n+1})=\prod_{m=0}^{\infty}\frac{1}{(1-q^{2m+2})}$$

$$\sum_{n=-\infty}^{\infty}q^{n^2}z^n\left(1+\sum_{m=1}^{\infty}\frac{(-1)^m q^m z^{-m}}{(1-q^2)(1-q^4)\ldots(1-q^{2m})}\right). \quad \text{(13-2-18)}$$

Finally, in (13–2–2) we replace q by q^2 and z by $-qz^{-1}$. This yields the formula

$$\prod_{n=0}^{\infty}\frac{1}{(1+z^{-1}q^{2n+1})}=1+\sum_{m=1}^{\infty}\frac{(-1)^m q^m z^{-m}}{(1-q^2)(1-q^4)\ldots(1-q^{2m})}, \quad \text{(13-2-19)}$$

provided that $|z|>|q|$ as we have specified. Substituting (13–2–19) into (13–2–18), we find that

$$\prod_{n=0}^{\infty}(1+zq^{2n+1})=\prod_{m=0}^{\infty}\frac{1}{(1-q^{2m+2})(1+z^{-1}q^{2m+1})}\sum_{n=-\infty}^{\infty}q^{n^2}z^n.$$

Multiplying both sides of this equation by

$$\prod_{m=0}^{\infty}(1-q^{2m+2})(1+z^{-1}q^{2m+1}),$$

we obtain (13–2–12) for $|z|>|q|$ and $|q|<1$.

Now we may repeat the entire argument with z^{-1} replacing z. Since (13–2–12) is symmetric in z and z^{-1}, our final result will be the same. The accompanying conditions, however, will now be $|z^{-1}|>|q|$ and $z^{-1}\neq 0$.

Since $|q|<1$, at least one of the inequalities $|z|>|q|$ and $|z^{-1}|>|q|$ must hold. Hence, (13–2–12) holds, provided only that $z\neq 0$ and $|q|<1$. ■

EXERCISES

1. Prove that if $H(z)=\prod_{n=0}^{\infty}\frac{(1+azq^n)}{(1-zq^n)}$, then $(1-z)H(z)=(1+az)H(zq)$.

2. Using Exercise 1 and supposing that

$$H(z)=\sum_{n=0}^{\infty}R_n z^n,$$

prove that $R_0=1$ and $(1-q^n)R_n=(1+aq^{n-1})R_{n-1}$.

3. Using Exercise 2, prove that

$$R_n = \frac{(1+a)(1+aq)\ldots(1+aq^{n-1})}{(1-q)(1-q^2)\ldots(1-q^n)},$$

and from this deduce that

$$1+\sum_{n=1}^{\infty}\frac{(1+a)(1+aq)\ldots(1+aq^{n-1})z^n}{(1-q)(1-q^2)\ldots(1-q^n)} = \prod_{n=0}^{\infty}\frac{(1+azq^n)}{(1-zq^n)},$$

provided that $|z| < 1$ and $|q| < 1$.

4. Use Exercise 3 to prove the following product-series identity:

$$1+\sum_{n=1}^{\infty}\frac{(b+a)(b+aq)\ldots(b+aq^{n-1})z^n}{(1-q)(1-q^2)\ldots(1-q^n)} = \prod_{n=0}^{\infty}\frac{(1+azq^n)}{(1-bzq^n)},$$

provided that $|zb| < 1$ and $|q| < 1$.

5. Deduce Theorem 13–7 from the result in Exercise 4.

In Exercises 6 through 9, you will need Theorem 13–7. Assume that $|q| < 1$ and $|a| < 1$.

6. Prove that

$$\sum_{n=0}^{\infty}\frac{q^{n(n+1)/2}(1-a)(1-aq)\ldots(1-aq^{n-1})}{(1-q)(1-q^2)\ldots(1-q^n)} = \prod_{m=0}^{\infty}(1-aq^m)$$

$$\sum_{n=0}^{\infty}\frac{q^{n(n+1)/2}}{(1-q)\ldots(1-q^n)}\sum_{r=0}^{\infty}\frac{a^r q^{rn}}{(1-q)\ldots(1-q^r)}.$$

7. Use Exercise 6 to prove that

$$\sum_{n=0}^{\infty}\frac{q^{n(n+1)/2}(1-a)(1-aq)\ldots(1-aq^{n-1})}{(1-q)\ldots(1-q^n)} = \prod_{m=0}^{\infty}(1-aq^m)$$

$$\sum_{r=0}^{\infty}\frac{a^r}{(1-q)\ldots(1-q^r)}\prod_{s=0}^{\infty}(1+q^{r+s+1}).$$

8. Use Exercise 7 to prove that

$$\sum_{n=0}^{\infty} \frac{q^{n(n+1)/2}(1-a)(1-aq)\ldots(1-aq^{n-1})}{(1-q)\ldots(1-q^n)} = \prod_{m=0}^{\infty}(1-aq^m)$$

$$\prod_{s=1}^{\infty}(1+q^s)\sum_{r=0}^{\infty}\frac{a^r}{(1-q^2)(1-q^4)\ldots(1-q^{2r})}.$$

9. Use Exercise 8 to prove that

$$\sum_{n=0}^{\infty} \frac{q^{n(n+1)/2}(1-a)(1-aq)\ldots(1-aq^{n-1})}{(1-q)\ldots(1-q^n)} =$$

$$\prod_{m=0}^{\infty}(1-aq^{2m+1})(1+q^{m+1}).$$

10. Let $b(n)$ denote the number of partitions of n into non-negative powers of 2. (Thus $b(5)=4$ since $5=4+1=2+2+1=2+1+1+1=1+1+1+1+1$.) Prove that

$$\sum_{n=0}^{\infty} b(n)q^n = \prod_{n=0}^{\infty}(1-q^{2^n})^{-1}.$$

11. Using the result in Exercise 10, prove the following three identities:

$$b(2n+1) = b(2n)$$

$$b(2n) = b(2n-1) + b(n)$$

$$b(n) \equiv 0 \pmod 2 \text{ for each } n > 1.$$

CHAPTER 14

PARTITION IDENTITIES

14–1 HISTORY AND INTRODUCTION

The study of partition identities originated with Euler in 1748 in his book *Introductio in Analysin Infinitorum.* Among his important contributions are Theorem 12–1 and his Pentagonal Number Theorem, proved in Section 14–2.

Problem 12–2 of Section 12–4 led us to conjecture that the following proposition is true.

THEOREM 14–1:

$$D_2(n) = p(S_2,n), \tag{14-1-1}$$

where $D_2(n)$ denotes the number of partitions of n in which any two summands differ by at least 2, and where $p(S_2,n)$ denotes the number of partitions of n into parts congruent to 1 or 4 modulo 5.

The proof we shall give is the simplest known proof of (14–1–1); it is nevertheless troublesome, and its discovery is a tribute to the ingenuity of both Rogers and Ramanujan. In 1894, L. J. Rogers proved a formula (equation (14–4–1)) that is a disguised form of the equation

$$\sum_{n=0}^{\infty} D_2(n)q^n = \sum_{n=0}^{\infty} p(S_2,n)q^n. \tag{14-1-2}$$

Note that this is merely (14–1–1) stated in terms of generating functions. Formula (14–4–1) is buried in the middle of Rogers's magnificent but long paper, "Second memoir on the expansion of infinite products." Apparently, Rogers's contemporaries were exhausted after reading the first memoir, because this second work, containing his proof of (14–4–1), drifted into obscurity.

In 1916, the famous Indian number theorist, S. Ramanujan, unaware of Rogers's work, discovered the same identity (14–4–1). Unable to prove the formula, Ramanujan referred it to G. H. Hardy at Cambridge, but neither he nor any of his English colleagues could find a proof. Thus, in 1916, P. A. MacMahon, in his monumental two-volume work, *Combinatory Analysis*, discusses Ramanujan's *un*proved result. Some time later, Ramanujan was thumbing through an old volume of the *Proceedings of the London Mathematical Society* when he suddenly happened onto Rogers's proof. In the ensuing decade, Ramanujan, Rogers, G. N. Watson, and others actively studied this identity and similar results.

We shall give the proof that Rogers and Ramanujan established jointly; for convenience, we shall at the same time prove Theorem 14–2, often referred to as the second Rogers-Ramanujan identity.

Theorem 14–2:

$$D_2'(n) = p(T_2, n), \tag{14-1-3}$$

where $D_2'(n)$ is the number of partitions of n in which any two summands differ by at least 2 and all summands are greater than 1; $p(T_2,n)$ is the number of partitions of n into parts congruent to 2 or 3 modulo 5.

Finally, no discussion of partition identities would be complete without mention of I. Schur. Schur, in Germany, isolated from English mathematicians during World War I, independently discovered (14–1–1) and (14–1–3) in 1917. Perhaps following the procedure outlined in Section 12–4, Schur studied $D_3(n)$, and in 1926 he proved the following related result.

Theorem 14–3:

$$\bar{D}_3(n) = p(S_3, n). \tag{14-1-4}$$

Here, $\bar{D}_3(n)$ denotes the number of partitions of n in which the difference between parts is at least 3, and in which no consecutive multiples of 3 are allowed; $p(S_3,n)$ is the number of partitions of n into parts congruent to 1 or 5 modulo 6.

We shall prove (14–1–4) in Section 14–5.

14–2 EULER'S PENTAGONAL NUMBER THEOREM

We obtain one of the most interesting corollaries of Jacobi's Triple Product Identity by replacing q by $q^{3/2}$, and z by $-q^{-1/2}$. This yields Euler's Pentagonal Number Theorem:

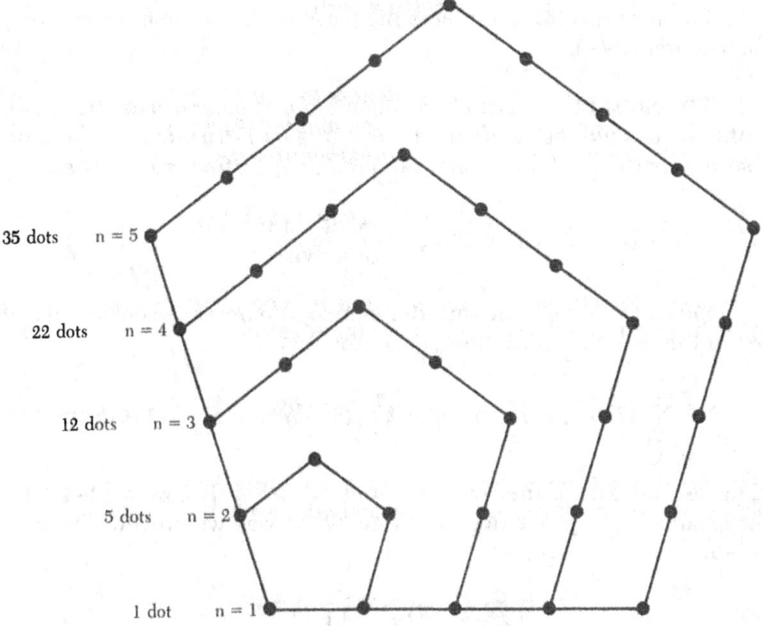

Figure 14-1 Pentagonal numbers.

THEOREM 14-4:

$$\sum_{n=-\infty}^{\infty} (-1)^n q^{n(3n-1)/2} = \prod_{n=1}^{\infty} (1-q^n), \textit{ where } |q| < 1.$$

REMARK: In Figure 14-1, we see that the number of dots inside and on the nth pentagon is $n(3n-1)/2$. Thus, the numbers 1, 5, 12, . . ., $n(3n-1)/2$, . . . are called *pentagonal numbers.*

PROOF: By Theorem 13-8, with q replaced by $q^{3/2}$, and z replaced by $-q^{-1/2}$, we have the relations

$$\sum_{n=-\infty}^{\infty} (-1)^n q^{n(3n-1)/2} = \prod_{n=0}^{\infty} (1-q^{3n+3})(1-q^{3n+1})(1-q^{3n+2})$$

$$= \prod_{n=1}^{\infty} (1-q^n),$$

because the three sequences $\{3n+3\}_{n=0}^{\infty}$, $\{3n+1\}_{n=0}^{\infty}$, and $\{3n+2\}_{n=0}^{\infty}$ contain all the positive integers with no repetitions. ■

The next result is an interesting partition-theoretic interpretation of Theorem 14–4.

THEOREM 14–5: *Let $D^e(n)$ denote the number of partitions of n into an even number of distinct parts, and let $D^o(n)$ denote the number of partitions of n into an odd number of distinct parts. Then*

$$D^e(n) - D^o(n) = \begin{cases} (-1)^j & \text{if } n = j(3j \pm 1)/2, \\ 0 & \text{otherwise.} \end{cases}$$

PROOF: We obtain this theorem by comparing coefficients of q^n in the first and third members of the relation

$$1 + \sum_{n=1}^{\infty} (D^e(n) - D^o(n))q^n = \prod_{m=1}^{\infty} (1 - q^m) = \sum_{j=-\infty}^{\infty} (-1)^j q^{j(3j-1)/2}.$$

The second part of the two equalities is merely Theorem 14–4. The first part we can prove in almost exactly the way we proved Theorem 13–5:

$$1 + \sum_{n=1}^{\infty} D_1(n)q^n = \prod_{n=1}^{\infty} (1 + q^n).$$

The only change is that, when we obtain the coefficient of q^n in the series expansion of $\prod_{j=1}^{m} (1 - q^j)$, each partition of n into distinct parts contributes +1 or −1 to the coefficient, depending on whether the number of parts in the partition is even or odd respectively. ■

EXERCISES

1. Prove that $\sum_{n=-\infty}^{\infty} (-1)^n q^{n^2} = \prod_{n=1}^{\infty} \frac{(1 - q^n)}{(1 + q^n)}$. [Hint: Set $z = -1$ in Theorem 13–8.]

2. Prove that $\sum_{n=-\infty}^{\infty} q^{n^2+n} = 2 \prod_{n=1}^{\infty} \frac{(1 - q^{4n})}{(1 - q^{4n-2})}$. [Hint: Set $z = q$ in Theorem 13–8.]

3. Prove that $\prod_{n=0}^{\infty} \{(1 - q^{2kn+k-h})(1 - q^{2kn+k+h})(1 - q^{2kn+2k})\} =$ $\sum_{n=-\infty}^{\infty} (-1)^n q^{kn^2+hn}$.

14-3 THE ROGERS-RAMANUJAN IDENTITIES

This section contains a five-part proof of Theorems 14–1 and 14–2. First we establish certain important equations (recurrence formulae) for the partition functions under consideration. Then we show that these equations fully determine our partition functions. After this, we consider the related generating functions, and from the effect of the recurrence formulae on the generating functions, we are able to deduce the Rogers-Ramanujan identities.

RECURRENCE FORMULAE FOR THE δ-PARTITIONS

To begin, we must examine partitions of the type enumerated by $D_2(n)$; let $\delta_i(m,n)$ denote the number of partitions of n into m distinct parts where 1 appears at most i times, and in which any two summands differ by at least 2. Our first objective is to verify the two recurrence equations

$$\delta_1(m,n) = \delta_0(m,n) + \delta_0(m-1,n-m), \tag{14-3-1}$$

and

$$\delta_0(m,n) = \delta_1(m,n-m). \tag{14-3-2}$$

To verify (14–3–1), we divide the partitions enumerated by $\delta_1(m,n)$ into two classes; those in the first class contain 1 as a summand, and those in the second class do not contain 1. The elements of the second class are exactly the partitions enumerated by $\delta_0(m,n)$.

Let us transform all the partitions in the first class by deleting the summand 1 from each, and then subtracting 1 from each of the remaining summands. This transformation leaves each partition in the first class with one less part, and it reduces the number being partitioned to $n-m$. Furthermore, since each partition originally contained 1 as a summand, all of the other parts must have been larger than 2 (by virtue of the proscription of consecutive integers in the original partition). Thus, after our transformation, all parts are larger than 1. We have therefore obtained partitions of the type enumerated by $\delta_0(m-1,n-m)$.

We can reverse the preceding transformation as follows. Given any partition of $n-m$ into $m-1$ distinct parts, each larger than 1, and with no consecutive integers appearing as summands, we may add 1 to every part and insert 1 as a summand to produce the elements of the first class. In this way, we have established that there are exactly $\delta_0(m-1,n-m)$ elements of the first class. Since the total number of elements in both classes equals $\delta_1(m,n)$, we see that (14–3–1) is established.

To verify (14–3–2), we merely apply the transformation to all partitions enumerated by $\delta_0(n,m)$. Then, as above, we establish that there are $\delta_1(m,n-m)$ partitions being counted.

Before we proceed toward our goal of proving (14–1–1) and (14–1–3), let us clarify the previous reasoning by examining a special case.

Example 14–1: Consider (14–3–1) and (14–3–2) in the case $m=3$, $n=15$. The following tables enumerate all the partitions involved, showing the partitions that correspond under the transformations.

TABLE 14–1: ILLUSTRATION OF (14–3–1)
$\delta_1(3,15)=\delta_0(3,15)+\delta_0(2,12)$

$\delta_1(3,15)=7$		$\delta_0(3,15)=3$		$\delta_0(2,12)=4$
$9+4+2$	$\longleftrightarrow$	$9+4+2$		
$8+5+2$	$\longleftrightarrow$	$8+5+2$		
$7+5+3$	$\longleftrightarrow$	$7+5+3$		
$11+3+1$	$\longleftrightarrow$		$\longleftrightarrow$	$10+2$
$10+4+1$	$\longleftrightarrow$		$\longleftrightarrow$	$9+3$
$9+5+1$	$\longleftrightarrow$		$\longleftrightarrow$	$8+4$
$8+6+1$	$\longleftrightarrow$		$\longleftrightarrow$	$7+5$

TABLE 14–2: ILLUSTRATION OF (14–3–2)
$\delta_0(3,15)=\delta_1(3,12)$

$\delta_0(3,15)=3$		$\delta_1(3,12)=3$
$9+4+2$	$\longleftrightarrow$	$8+3+1$
$8+5+2$	$\longleftrightarrow$	$7+4+1$
$7+5+3$	$\longleftrightarrow$	$6+4+2$

In studying (14–3–1) and (14–3–2), we tacitly assumed that neither class of partitions is empty. If one class *is* empty, we may easily validate (14–3–1) and (14–3–2) by the following definition:

$$\delta_1(m,n)=\delta_0(m,n)=\begin{cases}1 & \text{if } m=n=0,\\ 0 & \text{if either } m \text{ or } n \text{ is nonpositive and not both are zero.}\end{cases} \tag{14-3-3}$$

The equation $\delta_1(0,0) = \delta_0(0,0) = 1$ accounts for the "empty" partition of zero.

UNIQUENESS OF FUNCTIONS SATISFYING RECURRENCE FORMULAE

It is now important to show that (14–3–1), (14–3–2), and (14–3–3) completely define $\delta_1(m,n)$ and $\delta_0(m,n)$. By "completely define," we mean that if $c_1(m,n)$ and $c_0(m,n)$ are arbitrary functions defined for all integral values of m and n, and

$$c_1(m,n) = c_0(m,n) + c_0(m-1,n-m), \tag{14-3-1$'$}$$

$$c_0(m,n) = c_1(m,n-m), \tag{14-3-2$'$}$$

$$c_1(m,n) = c_0(m,n) = \begin{cases} 1 & \text{if } m = n = 0 \\ 0 & \text{if either } m \text{ or } n \text{ is nonpositive} \\ & \text{and not both are zero,} \end{cases} \tag{14-3-3$'$}$$

then $\delta_1(m,n) = c_1(m,n)$ and $\delta_0(m,n) = c_0(m,n)$.

This last assertion is proved by mathematical induction applied to n. By (14–3–3) and (14–3–3)$'$, $\delta_1(m,n) = c_1(m,n)$ and $\delta_0(m,n) = c_0(m,n)$ for all $n \leq 0$. Assume that $\delta_1(m,n) = c_1(m,n)$ and $\delta_0(m,n) = c_0(m,n)$ for all $n \leq k$, where $k \geq 0$. Then, by (14–3–2)$'$,

$$c_0(m,k+1) = c_1(m,k+1-m).$$

If $m \leq 0$, then by (14–3–3) and (14–3–3)$'$, $c_1(m,k+1-m) = \delta_1(m,k+1-m)$; if $m > 0$, then $k+1-m \leq k$, and $c_1(m,k+1-m) = \delta_1(m,k+1-m)$ by the induction hypothesis. In any case,

$$\begin{aligned} c_0(m,k+1) &= c_1(m,k+1-m) \\ &= \delta_1(m,k+1-m) \\ &= \delta_0(m,k+1), \end{aligned}$$

by (14–3–2).

Now

$$\begin{aligned} c_1(m,k+1) &= c_0(m,k+1) + c_0(m-1,k+1-m) \quad \text{(by (14–3–1)$'$)} \\ &= \delta_0(m,k+1) + \delta_0(m-1,k+1-m) \\ &= \delta_1(m,k+1). \end{aligned}$$

Thus we have shown by mathematical induction that $\delta_1(m,n) = c_1(m,n)$ and $\delta_0(m,n) = c_0(m,n)$, for all integers m and n.

GENERATING FUNCTIONS FOR THE δ-PARTITION FUNCTIONS

We define

$$G_1(x;q) = \sum_{M=0}^{\infty} \sum_{N=0}^{\infty} \delta_1(M,N) x^M q^N, \tag{14-3-4}$$

and

$$G_0(x;q) = \sum_{M=0}^{\infty} \sum_{N=0}^{\infty} \delta_0(M,N) x^M q^N. \tag{14-3-5}$$

The interested reader may refer to Appendix C, where it is shown that such series converge for $|q| < 1$ and $|x| < |q|^{-1}$.

We now use (14–3–1), (14–3–2), and (14–3–3) to study $G_1(x;q)$ and $G_2(x;q)$. The definitions (14–3–4) and (14–3–5) lead us to three important identities:

$$\begin{aligned}
G_1(x;q) &= \sum_{M=0}^{\infty} \sum_{N=0}^{\infty} \delta_1(M,N) x^M q^N \\
&= \sum_{M=0}^{\infty} \sum_{N=0}^{\infty} (\delta_0(M,N) + \delta_0(M-1,N-M)) x^M q^N \\
&= \sum_{M=0}^{\infty} \sum_{N=0}^{\infty} \delta_0(M,N) x^M q^N + \sum_{M=0}^{\infty} \sum_{N=0}^{\infty} \delta_0(M-1,N-M) x^M q^N \\
&= G_0(x;q) + xq \sum_{M=0}^{\infty} \sum_{N=0}^{\infty} \delta_0(M-1,N-M)(xq)^{M-1} q^{N-M} \\
&= G_0(x;q) + xq \sum_{M=-1}^{\infty} \sum_{N=-M-1}^{\infty} \delta_0(M,N)(xq)^M q^N \\
&= G_0(x;q) + xqG_0(xq;q).
\end{aligned} \tag{14-3-6}$$

Similarly,

$$\begin{aligned}
G_0(x;q) &= \sum_{M=0}^{\infty} \sum_{N=0}^{\infty} \delta_0(M,N) x^M q^N \\
&= \sum_{M=0}^{\infty} \sum_{N=0}^{\infty} \delta_1(M,N-M) x^M q^N
\end{aligned}$$

$$= \sum_{M=0}^{\infty} \sum_{N=-M}^{\infty} \delta_1(M,N) x^M q^{N+M}$$

$$= \sum_{M=0}^{\infty} \sum_{N=0}^{\infty} \delta_1(M,N) (xq)^M q^N$$

$$= G_1(xq;q). \tag{14-3-7}$$

Furthermore, (14–3–3) implies that

$$G_1(0;q) = G_0(0;q) = 1. \tag{14-3-8}$$

We can now turn the tables: we reverse our steps and deduce (14–3–1), (14–3–2), and (14–3–3) from (14–3–6), (14–3–7), and (14–3–8). Thus, if we can find any functions

$$g_1(x;q) = \sum_{N=0}^{\infty} \sum_{M=0}^{\infty} c_1(M,N) x^M q^N$$

and

$$g_0(x;q) = \sum_{N=0}^{\infty} \sum_{M=0}^{\infty} c_0(M,N) x^M q^N,$$

satisfying the conditions

$$g_1(x;q) = g_0(x;q) + xqg_0(xq;q), \tag*{(14-3-6)$'$}$$

$$g_0(x;q) = g_1(xq;q), \tag*{(14-3-7)$'$}$$

$$g_1(0;q) = g_0(0;q) = 1, \tag*{(14-3-8)$'$}$$

we may deduce that $c_1(M,N)$ and $c_0(M,N)$ satisfy (14–3–1)′, (14–3–2)′, and (14–3–3)′. Consequently, by what we established earlier, $c_1(M,N) = \delta_1(M,N)$ and $c_0(M,N) = \delta_0(M,N)$. Hence $g_1(x;q) = G_1(x;q)$ and $g_0(x;q) = G_0(x;q)$.

NEW FUNCTIONS

At this stage, then, we need some "new" functions of x and q that satisfy (14–3–6)′, (14–3–7)′, and (14–3–8)′. The ingenuity of the pioneers Rogers and Ramanujan produced the following functions. For any integer i, and for $|q| < 1$ and $|x| < |q|^{-1}$,

$$f_i(x;q)=\sum_{n=0}^{\infty}\frac{(-1)^n x^{2n} q^{\frac{1}{2}n(5n+3)-in}\,(1-x^{i+1}q^{(2n+1)(i+1)})}{(1-q)(1-q^2)\ldots(1-q^n)\prod_{j=n+1}^{\infty}(1-xq^j)}. \tag{14-3-9}$$

When $n=0$, the summand is by convention just

$$(1-x^{i+1}q^{i+1})\Big/\prod_{j=1}^{\infty}(1-xq^j),$$

since for $n=0$, the empty product $(1-q)(1-q^2)\ldots(1-q^n)$ is defined as 1. When $i=-1$,

$$1-x^{i+1}q^{(2n+1)(i+1)}=1-x^0q^0=1-1=0.$$

Thus, each term in the series for $f_{-1}(x;q)$ is zero. Therefore,

$$f_{-1}(x;q)=0. \tag{14-3-10}$$

Now

$$f_i(x;q)-f_{i-1}(x;q)$$

$$=\sum_{n=0}^{\infty}\frac{(-1)^n x^{2n} q^{\frac{1}{2}n(5n+3)}(q^{-in}-x^{i+1}q^{(2n+1)(i+1)-in}-q^{-(i-1)n}+x^iq^{(2n+1)i-(i-1)n})}{(1-q)(1-q^2)\ldots(1-q^n)\prod_{j=n+1}^{\infty}(1-xq^j)}$$

$$=\sum_{n=0}^{\infty}\frac{(-1)^n x^{2n} q^{\frac{1}{2}n(5n+3)}(q^{-in}(1-q^n)+x^iq^{(2n+1)i-(i-1)n}(1-xq^{n+1}))}{(1-q)(1-q^2)\ldots(1-q^n)\prod_{j=n+1}^{\infty}(1-xq^j)}$$

$$=\sum_{n=0}^{\infty}\frac{(-1)^n x^{2n} q^{\frac{1}{2}n(5n+3)}\,q^{-in}(1-q^n)}{(1-q)(1-q^2)\ldots(1-q^n)\prod_{j=n+1}^{\infty}(1-xq^j)}$$

$$+\sum_{n=0}^{\infty}\frac{(-1)^n x^{2n} q^{\frac{1}{2}n(5n+3)}x^iq^{(2n+1)i-(i-1)n}(1-xq^{n+1})}{(1-q)(1-q^2)\ldots(1-q^n)\prod_{j=n+1}^{\infty}(1-xq^j)}$$

We note that the first term in the first sum of the last expression is simply $(1-q^0)\Big/\prod_{j=1}^{\infty}(1-xq^j)=0$. Thus, we may really treat the first sum as running from $n=1$ to ∞. Replacing n by $n+1$ in the first sum, we find that

$$f_i(x;q) - f_{i-1}(x;q)$$

$$= \sum_{n=0}^{\infty} \frac{(-1)^{n+1} x^{2n+2} q^{\frac{1}{2}(n+1)(5n+8)} q^{-in-i}}{(1-q)(1-q^2)\ldots(1-q^n)\prod_{j=n+2}^{\infty}(1-xq^j)}$$

$$+ \sum_{n=0}^{\infty} \frac{(-1)^n x^{2n} q^{\frac{1}{2}n(5n+3)} x^i q^{(2n+1)i-(i-1)n}}{(1-q)(1-q^2)\ldots(1-q^n)\prod_{j=n+2}^{\infty}(1-xq^j)}$$

$$= x^i q^i \sum_{n=0}^{\infty} \frac{(-1)^n x^{2n} q^{\frac{1}{2}n(5n+3)+in+n}(1 - x^{2-i} q^{(2n+2)(2-i)})}{(1-q)(1-q^2)\ldots(1-q^n)\prod_{j=n+1}^{\infty}(1-xqq^j)}$$

$$= x^i q^i \sum_{n=0}^{\infty} \frac{(-1)^n (xq)^{2n} q^{\frac{1}{2}n(5n+3)-(1-i)n}(1-(xq)^{2-i} q^{(2n+1)(2-i)})}{(1-q)(1-q^2)\ldots(1-q^n)\prod_{j=n+1}^{\infty}(1-xqq^{j+1})}$$

$$= x^i q^i f_{1-i}(xq;q). \tag{14-3-11}$$

Now, if we set $i=1$, (14–3–11) implies that

$$f_1(x;q) = f_0(x;q) + xqf_0(xq;q). \tag{14-3-12}$$

If we set $i=0$ and recall that (14–3–10) asserts that $f_{-1}(x;q)=0$, we see that by (14–3–11)

$$f_0(x;q) = f_1(xq;q). \tag{14-3-13}$$

Finally, if we set $x=0$ in (14–3–9), we see that

$$f_0(0;q) = f_1(0;q) = 1. \tag{14-3-14}$$

In Appendix C it is shown that $f_i(x;q)$ has an expansion of the form

$$f_i(x;q) = \sum_{N=0}^{\infty}\sum_{M=0}^{\infty} c_i(M,N)x^M q^N. \tag{14-3-15}$$

Equations (14–3–12), (14–3–13), and (14–3–14) are simply (14–3–6)′, (14–3–7)′, and (14–3–8)′ with f replacing g. Thus we may conclude that $f_1(x;q) = G_1(x;q)$ and $f_0(x;q) = G_0(x;q)$.

THE "COUP DE GRACE"

Clearly,

$$\sum_{M=0}^{\infty} \delta_1(M,N) = D_2(N)$$

and

$$\sum_{M=0}^{\infty} \delta_0(M,N) = D_2'(N).$$

Thus,

$$\begin{aligned}
\sum_{N=0}^{\infty} D_2(N)\,q^N &= \sum_{M=0}^{\infty}\sum_{N=0}^{\infty} \delta_1(M,N)\,q^N \\
&= G_1(1;q) \\
&= f_1(1;q) \\
&= \frac{\sum_{n=0}^{\infty} (-1)^n q^{\frac{1}{2}n(5n+1)}\,(1-q^{4n+2})}{\prod_{j=1}^{\infty}(1-q^j)} \\
&= \frac{\sum_{n=0}^{\infty} (-1)^n q^{\frac{1}{2}n(5n+1)} - \sum_{n=0}^{\infty} (-1)^n q^{\frac{1}{2}n(5n+9)+2}}{\prod_{j=1}^{\infty}(1-q^j)}
\end{aligned}$$

If in the second sum of the last expression we replace n by $-n-1$, we find that

$$\begin{aligned}
\sum_{N=0}^{\infty} D_2(N)q^N &= \frac{\sum_{n=0}^{\infty} (-1)^n q^{\frac{1}{2}n(5n+1)} - \sum_{n=-1}^{-\infty} (-1)^{-n-1} q^{\frac{1}{2}(-n-1)(-5n+4)+2}}{\prod_{j=1}^{\infty}(1-q^j)} \\
&= \frac{\sum_{n=0}^{\infty} (-1)^n q^{\frac{1}{2}n(5n+1)} + \sum_{n=-1}^{-\infty} (-1)^n q^{\frac{1}{2}n(5n+1)}}{\prod_{j=1}^{\infty}(1-q^j)}
\end{aligned}$$

$$= \frac{\sum_{n=-\infty}^{\infty} (-1)^n q^{\frac{1}{2}n(5n+1)}}{\prod_{j=1}^{\infty} (1-q^j)}. \tag{14-3-16}$$

By Jacobi's Triple Product Identity (Theorem 13–8), with q replaced by $q^{5/2}$ and z by $-q^{1/2}$,

$$\sum_{n=-\infty}^{\infty} (-1)^n q^{\frac{1}{2}n(5n+1)} = \prod_{n=0}^{\infty} (1-q^{5n+5})(1-q^{5n+3})(1-q^{5n+2}). \tag{14-3-17}$$

Since the five sequences $\{5n+5\}_{n=0}^{\infty}$, $\{5n+1\}_{n=0}^{\infty}$, $\{5n+2\}_{n=0}^{\infty}$, $\{5n+3\}_{n=0}^{\infty}$, and $\{5n+4\}_{n=0}^{\infty}$ contain all the positive integers without duplications,

$$\prod_{j=1}^{\infty} (1-q^j) = \prod_{n=0}^{\infty} (1-q^{5n+5})(1-q^{5n+1})(1-q^{5n+2})(1-q^{5n+3})(1-q^{5n+4}). \tag{14-3-18}$$

Substituting (14–3–17) and (14–3–18) into (14–3–16), we have the formula

$$\sum_{N=0}^{\infty} D_2(N)q^N = \frac{\prod_{n=0}^{\infty} (1-q^{5n+5})(1-q^{5n+2})(1-q^{5n+3})}{\prod_{n=0}^{\infty} (1-q^{5n+5})(1-q^{5n+1})(1-q^{5n+2})(1-q^{5n+3})(1-q^{5n+4})}$$

$$= \prod_{n=0}^{\infty} \frac{1}{(1-q^{5n+1})(1-q^{5n+4})}$$

$$= \prod_{j=1}^{\infty} \frac{1}{(1-q^{5j-4})(1-q^{5j-1})}$$

$$= \sum_{N=0}^{\infty} p(S_2,N)q^N, \tag{14-3-19}$$

by (13–1–2).

Comparing coefficients on both sides of (14–3–19), we conclude that $D_2(N) = p(S_2,N)$, as was asserted in (14–1–1).

We have already accomplished enough to establish easily (14–1–3), the second Rogers-Ramanujan identity. Since it is similarly derived from our formulae for $f_i(x;q)$, we shall only outline the procedure.

$$\sum_{N=0}^{\infty} D_2'(N)q^n = G_0(1;q)$$

$$= f_0(1;q)$$

$$= \frac{\sum_{n=-\infty}^{\infty} (-1)^n q^{\frac{1}{2}n(5n+3)}}{\prod_{j=1}^{\infty} (1-q^j)}$$

$$= \frac{\prod_{n=0}^{\infty} (1-q^{5n+5})(1-q^{5n+1})(1-q^{5n+4})}{\prod_{n=0}^{\infty} (1-q^{5n+5})(1-q^{5n+1})(1-q^{5n+2})(1-q^{5n+3})(1-q^{5n+4})}$$

$$= \prod_{n=0}^{\infty} \frac{1}{(1-q^{5n+2})(1-q^{5n+3})}$$

$$= \sum_{N=0}^{\infty} p(T_2,N)\, q^N. \qquad (14\text{-}3\text{-}20)$$

Comparing coefficients on both sides of (14–3–20), we conclude that $D_2'(N) = p(T_2,N)$, as was asserted in (14–1–3). ■

14–4 SERIES AND PRODUCT IDENTITIES

In this section, we shall give the two identities originally discovered by Rogers.

THEOREM 14–6: *If $|q| < 1$, then*

$$1 + \sum_{n=1}^{\infty} \frac{q^{n^2}}{(1-q)(1-q^2)\ldots(1-q^n)} = \prod_{n=0}^{\infty} \frac{1}{(1-q^{5n+1})(1-q^{5n+4})}, \qquad (14\text{-}4\text{-}1)$$

and

$$1 + \sum_{n=1}^{\infty} \frac{q^{n^2+n}}{(1-q)(1-q^2)\ldots(1-q^n)} = \prod_{n=0}^{\infty} \frac{1}{(1-q^{5n+3})(1-q^{5n+4})}. \qquad (14\text{-}4\text{-}2)$$

PROOF: We start with the substitution of (14–3–13) into (14–3–12), which yields

$$f_1(x;q) = f_1(xq;q) + xqf_1(xq^2;q). \qquad (14\text{-}4\text{-}3)$$

Let

$$f_1(x;q)=\sum_{n=0}^{\infty} B_n x^n, \tag{14-4-4}$$

where B_n depends on q. Substituting this series into (14–4–3), we find that

$$\sum_{n=0}^{\infty} B_n x^n = \sum_{n=0}^{\infty} B_n q^n x^n + \sum_{n=0}^{\infty} B_n q^{2n+1} x^{n+1}. \tag{14-4-5}$$

Comparing coefficients of x^n on both sides of this equation, we find that

$$B_n = B_n q^n + q^{2n-1} B_{n-1};$$

since $f_1(0;q)=1$, we see that $B_0=1$. Therefore,

$$\begin{aligned} B_n &= \frac{q^{2n-1}}{1-q^n} B_{n-1} \\ &\ldots \\ &= \frac{q^{2n-1}}{1-q^n}\frac{q^{2n-3}}{1-q^{n-1}}\cdots\frac{q}{1-q} B_0 \\ &= \frac{q^{(2n-1)+(2n-3)+\ldots+3+1}}{(1-q)\ldots(1-q^n)} \\ &= \frac{q^{n^2}}{(1-q)\ldots(1-q^n)}. \end{aligned}$$

Hence,

$$f_1(x;q)=1+\sum_{n=1}^{\infty}\frac{q^{n^2}x^n}{(1-q)(1-q^2)\ldots(1-q^n)}. \tag{14-4-6}$$

Recalling from (14–3–19) that

$$f_1(1;q)=\prod_{n=0}^{\infty}\frac{1}{(1-q^{5n+1})(1-q^{5n+4})},$$

we now obtain (14–4–1) by setting $x=1$ in (14–4–6). To prove (14–4–2), we note from (14–3–20) that

$$\begin{aligned} \prod_{n=0}^{\infty}\frac{1}{(1-q^{5n+2})(1-q^{5n+3})} &= f_0(1;q) \\ &= f_1(q;q) \quad \text{(by (14–3–13))} \end{aligned}$$

$$= 1 + \sum_{n=1}^{\infty} \frac{q^{n^2+n}}{(1-q)(1-q^2)\dots(1-q^n)},$$

by (14–4–6). ∎

14–5 SCHUR'S THEOREM

Our object here is to prove the result of I. Schur discussed in Section 14–1. The proof is easier than the proof of the Rogers-Ramanujan identities. We shall begin as in Section 14–3; however, we shall never need to pull a rabbit like (14–3–9) out of the hat. First we must prove a result, called Abel's Lemma, from the theory of infinite series.

THEOREM 14–7: *If* $\lim_{n\to\infty} a_n = L$, *then*

$$\lim_{x\to 1^-} (1-x) \sum_{n=0}^{\infty} a_n x^n = L.$$

PROOF: The comparison test with the series $\sum_{n=0}^{\infty} (|L|+1)x^n$ shows that $\sum_{n=0}^{\infty} a_n x^n$ actually converges for $|x| < 1$.

Recall that the statement "$\lim_{n\to\infty} a_n = L$" means that for each $\epsilon > 0$, there exists an N such that $|a_n - L| < \epsilon/2$ whenever $n \geq N$. Thus,

$$\left|(1-x)\sum_{n=0}^{\infty} a_n x^n - L\right| = \left|(1-x)\sum_{n=0}^{\infty} a_n x^n - (1-x)L\sum_{n=0}^{\infty} x^n\right|$$

$$\left(\text{since } \frac{1}{1-x} = \sum_{n=0}^{\infty} x^n\right)$$

$$= \left|(1-x)\sum_{n=0}^{\infty} (a_n - L)x^n\right|$$

$$\leq \left|(1-x)\sum_{n=0}^{N} (a_n - L)x^n\right|$$

$$+ \left|(1-x)\sum_{n=N+1}^{\infty} (a_n - L)x^n\right|.$$

Now let M be the largest of the numbers $|a_0-L|, |a_1-L|, |a_2-L|, \ldots, |a_N-L|$. Then, for $0<x<1$,

$$\left|(1-x)\sum_{n=0}^{\infty} a_n x^n - L\right| \le (1-x)M(N+1) + (1-x)\frac{\epsilon}{2}\frac{x^{N+1}}{1-x}$$

$$< (1-x)M(N+1) + \frac{\epsilon}{2};$$

this last expression is less than ϵ, provided that

$$1-\frac{\epsilon}{2M(N+1)} < x < 1.$$

Thus we see that

$$\lim_{x\to 1^-} (1-x)\sum_{n=0}^{\infty} a_n x^n = L. \quad \blacksquare$$

With Abel's Lemma among our tools, we are now prepared to prove Schur's Theorem.

THEOREM 14–8 (I. Schur): $\bar{D}_3(n) = p(S_3,n)$.

PROOF: Let $\Delta_j(m,n)$ denote the number of partitions of n into m distinct parts, each greater than j, such that the difference between any two parts is at least 3 and such that no consecutive multiples of 3 appear as parts.

We must establish the identities

$$\Delta_0(m,n) = \Delta_1(m,n) + \Delta_0(m-1,n-3m+2), \quad (14\text{-}5\text{-}1)$$

$$\Delta_1(m,n) = \Delta_2(m,n) + \Delta_1(m-1,n-3m+1), \quad (14\text{-}5\text{-}2)$$

$$\Delta_2(m,n) = \Delta_3(m,n) + \Delta_3(m-1,n-3m), \quad (14\text{-}5\text{-}3)$$

and

$$\Delta_3(m,n) = \Delta_0(m,n-3m). \quad (14\text{-}5\text{-}4)$$

Since the proofs of these equations are similar, all resembling the proofs of (14–3–1) and (14–3–2), we shall give a detailed proof only of (14–5–1).

To verify (14–5–1), we divide the partitions enumerated by $\Delta_0(m,n)$ into two classes, those in the first class containing 1 as a summand, and those in the second class not containing 1 as a summand. The elements of the second class are exactly the partitions enumerated by $\Delta_1(m,n)$.

Let us transform all the partitions in the first class by deleting the summand 1 from each, and then subtracting 3 from each of the remaining summands. This transformation leaves each partition in the first class with one less part, and it reduces the number being partitioned to $n-3(m-1)-1=n-3m+2$. Furthermore, since each partition originally contained 1 as a summand, all of the other parts had to be larger than 3; otherwise we should have had two parts differing by less than 3, a proscribed condition. Thus, after our transformation, all parts are at least as large as 1. We have therefore obtained partitions of the type enumerated by $\Delta_0(m-1,n-3m+2)$.

Conversely, given any partition of $n-3m+2$ into $m-1$ distinct parts such that the difference between any two parts is at least 3 and such that no consecutive multiples of 3 appear, we may add 3 to every part and insert 1 as a summand to produce the elements of the first class. In this way, we establish that there are exactly $\Delta_0(m-1, n-3m+2)$ elements of the first class. Since the total number of elements in both classes equals $\Delta_0(m,n)$, we see that (14–5–1) is established.

The proofs of (14–5–2), (14–5–3), and (14–5–4) are similar to the proof of (14–5–1); however, a slight complication arises in (14–5–3). Here, in the first class, the partitions do contain a 3, and therefore they must have all their other parts as large as 7, since 3 and 6 are consecutive multiples of 3.

In studying the four equations (14–5–1) to (14–5–4), we tacitly assumed that both classes of partitions were nonempty. If one class *is* empty, we may easily validate all four formulae by the following definition:

$$\Delta_0(m,n)=\Delta_1(m,n)=\Delta_2(m,n)=\Delta_3(m,n)=\begin{cases}1 \text{ if } m=n=0\\ 0 \text{ if either } m \text{ or } n \text{ is nonpositive and not both are zero.}\end{cases} \tag{14-5-5}$$

The top line of equation (14–5–5) accounts for the "empty" partition of zero.

For $i=0,1,2,3$ and $|q|<1$, $|x|<|q|^{-1}$, define

$$H_i(x;q)=\sum_{M=0}^{\infty}\sum_{N=0}^{\infty}\Delta_i(M,N)x^Mq^N.$$

We use (14–5–1) to (14–5–5) to study $H_i(x;q)$. By (14–5–1),

$$H_0(x;q)=\sum_{M=0}^{\infty}\sum_{N=0}^{\infty}\Delta_0(M,N)x^Mq^N$$

$$=\sum_{M=0}^{\infty}\sum_{N=0}^{\infty}(\Delta_1(M,N)+\Delta_0(M-1,N-3M+2))x^Mq^N$$

$$=\sum_{M=0}^{\infty}\sum_{N=0}^{\infty}\Delta_1(M,N)x^Mq^N$$

$$+\sum_{M=0}^{\infty}\sum_{N=0}^{\infty}\Delta_0(M-1,N-3M+2)x^Mq^N$$

$$=H_1(x;q)+xq\sum_{M=0}^{\infty}\sum_{N=0}^{\infty}\Delta_0(M-1,N-3M+2)(xq^3)^{M-1}q^{N-3M+2}$$

$$=H_1(x;q)+xq\sum_{M=-1}^{\infty}\sum_{N=-3M-1}^{\infty}\Delta_0(M,N)(xq^3)^Mq^N$$

$$=H_1(x;q)+xqH_0(xq^3;q). \tag{14-5-6}$$

In exactly the same way, (14–5–2), (14–5–3), and (14–5–4) imply that

$$H_1(x;q)=H_2(x;q)+xq^2H_1(xq^3;q), \tag{14-5-7}$$

$$H_2(x;q)=H_3(x;q)+xq^3H_3(xq^3;q), \tag{14-5-8}$$

$$H_3(x;q)=H_0(xq^3;q). \tag{14-5-9}$$

Substituting (14–5–9) into (14–5–8), we find that

$$H_2(x;q)=H_0(xq^3;q)+xq^3H_0(xq^6;q). \tag{14-5-10}$$

Solving (14–5–6) for $H_1(x;q)$, we find that

$$H_1(x;q)=H_0(x;q)-xqH_0(xq^3;q). \tag{14-5-11}$$

Substituting (14–5–10) and (14–5–11) into (14–5–7), we have the equation

$$H_0(x;q)-xqH_0(xq^3;q)=(H_0(xq^3;q)+xq^3H_0(xq^6;q))$$
$$+xq^2(H_0(xq^3;q)-xq^4H_0(xq^6;q)).$$

Simplifying this last equation, we have the relation

$$H_0(x;q) = (1+xq+xq^2)H_0(xq^3;q) + xq^3(1-xq^3)H_0(xq^6;q). \quad (14\text{-}5\text{-}12)$$

Define the function

$$h(x;q) = \frac{H_0(x;q)}{\prod_{n=0}^{\infty}(1-xq^{3n})}. \quad (14\text{-}5\text{-}13)$$

Dividing both sides of (14–5–12) by $\prod_{n=1}^{\infty}(1-xq^{3n})$, we find that

$$(1-x)h(x;q) = (1+xq+xq^2)h(xq^3;q) + xq^3h(xq^6;q). \quad (14\text{-}5\text{-}14)$$

Now let us consider $h(x;q)$ in the form

$$h(x;q) = \sum_{n=0}^{\infty} E_n x^n, \quad (14\text{-}5\text{-}15)$$

where E_n depends on q. Substituting (14–5–15) into (14–5–14), we obtain the formula

$$\sum_{n=0}^{\infty} E_n x^n - \sum_{n=0}^{\infty} E_n x^{n+1} = \sum_{n=0}^{\infty} E_n q^{3n} x^n + \sum_{n=0}^{\infty} E_n q^{3n+1} x^{n+1}$$

$$+ \sum_{n=0}^{\infty} E_n q^{3n+2} x^{n+1} + \sum_{n=0}^{\infty} E_n q^{6n+3} x^{n+1}. \quad (14\text{-}5\text{-}16)$$

By (14–5–15), (14–5–6), and (14–5–5), $h(0;q) = H_0(0;q) = \Delta_0(0,0) = 1$; thus $E_0 = 1$. Comparing coefficients of x^N on both sides of (14–5–16) for $N > 0$, we see that

$$E_N - E_{N-1} = q^{3N}E_N + q^{3N-2}E_{N-1} + q^{3N-1}E_{N-1} + q^{6N-3}E_{N-1},$$

and thus

$$(1-q^{3N})E_N = (1+q^{3N-2}+q^{3N-1}+q^{6N-3})E_{N-1}.$$

Hence,

$$E_N = \frac{(1+q^{3N-2})(1+q^{3N-1})}{1-q^{3N}} E_{N-1}$$

$$= \frac{(1+q^{3N-2})(1+q^{3N-1})(1+q^{3N-5})(1+q^{3N-4})}{(1-q^{3N})(1-q^{3N-3})} E_{N-2}$$

$$\cdots$$

$$= \prod_{j=1}^{N} \frac{(1+q^{3j-2})(1+q^{3j-1})}{(1-q^{3j})}. \quad (14\text{-}5\text{-}17)$$

Substituting (14–5–17) into (14–5–15) and comparing the result with (14–5–13), we find that

$$H_0(x;q) = \prod_{n=0}^{\infty} (1 - xq^{3n}) \sum_{m=0}^{\infty} x^m \prod_{j=1}^{m} \frac{(1+q^{3j-2})(1+q^{3j-1})}{(1-q^{3j})}. \qquad (14\text{-}5\text{-}18)$$

Using Abel's Lemma (Theorem 14–7), we shall now deduce Schur's Theorem from (14–5–18).

$$\begin{aligned}\sum_{N=0}^{\infty} \bar{D}_3(N) q^N &= \sum_{N=0}^{\infty} \sum_{M=0}^{\infty} \Delta_0(M,N) q^N \\ &= H_0(1;q) \\ &= \prod_{n=1}^{\infty} (1-q^{3n}) \lim_{x\to 1^-} (1-x) \sum_{m=0}^{\infty} x^m \prod_{j=1}^{m} \frac{(1+q^{3j-2})(1+q^{3j-1})}{(1-q^{3j})} \\ &= \prod_{n=1}^{\infty} (1-q^{3n}) \prod_{j=1}^{\infty} \frac{(1+q^{3j-2})(1+q^{3j-1})}{(1-q^{3j})} \\ &= \prod_{j=1}^{\infty} (1+q^{3j-2})(1+q^{3j-1}) \\ &= \sum_{n=0}^{\infty} Q_3(n) q^n \\ &= \sum_{n=0}^{\infty} p(S_3,n) q^n, \qquad (14\text{-}5\text{-}19)\end{aligned}$$

where the last equation follows from the proof of Theorem 13–6.

Comparing coefficients of q^n on both sides of (14–5–19), we obtain Theorem 14–8. ■

EXERCISES

Section 12–4 contains thirteen exercises on the formulation of conjectures concerning identities for various partition functions. Many of the possible conjectures can be proved using the techniques developed in Chapters 13 and 14. Partial results can be obtained for some of the more difficult exercises.

The notation is repeated from the Chapter 12 exercises.

1. Let $e_1(m,n)$ denote the number of partitions of n into m parts in which all parts differ by at least 2 and no consecutive even numbers appear as summands. Let $e_3(m,n)$ be the same, except that all parts must be as large as 3. Prove that $e_1(m,n) = e_3(m,n) + e_3(m-1,n-2m) + e_1(m-1, n-2m+1)$ and $e_3(m,n) = e_1(m,n-2m)$.

2. Let $\mathscr{E}_a(x) = \sum_{n=0}^{\infty} \sum_{m=0}^{\infty} e_a(m,n)x^m q^n$. Deduce from Exercise 1 that $\mathscr{E}_1(x) = \mathscr{E}_3(x) + xq\mathscr{E}_1(xq^2) + xq^2\mathscr{E}_3(xq^2)$ and $\mathscr{E}_3(x) = \mathscr{E}_1(xq^2)$.

3. Deduce from Exercise 2 that, for $|q| < 1$,

$$\mathscr{E}_1(x) = \sum_{n=0}^{\infty} \frac{x^n q^{n^2}(1+q)(1+q^3)\ldots(1+q^{2n-1})}{(1-q^2)(1-q^4)\ldots(1-q^{2n})},$$

$$\mathscr{E}_2(x) = \sum_{n=0}^{\infty} \frac{x^n q^{n^2+2n}(1+q)(1+q^3)\ldots(1+q^{2n-1})}{(1-q^2)(1-q^4)\ldots(1-q^{2n})}.$$

4. Use Exercise 3 and your conjectures from Exercises 1 and 3 of Chapter 12 to conjecture, for $\mathscr{E}_1(1)$ and $\mathscr{E}_3(1)$, series-product identities similar to those in Theorem 14–6.

5. Let $e_2(m,n)$ denote the number of partitions of n into m parts in which all parts differ by at least 2 and no consecutive odd numbers appear as summands. Let $e_4(m,n)$ be the same except that all parts must be at least 2. Prove that $e_2(m,n) = e_4(m,n) + e_4(m-1,n-2m+1)$ and $e_4(m,n) = e_2(m,n-2m) + e_4(m-1,n-2m)$.

6. Deduce from Exercise 5 that $\mathscr{E}_2(x) = \mathscr{E}_4(x) + xq\mathscr{E}_4(xq^2)$ and $\mathscr{E}_4(x) = \mathscr{E}_2(xq^2) + xq^2\mathscr{E}_4(xq^2)$.

7. Deduce from Exercise 6 that, for $|q| < 1$,

$$\mathscr{E}_4(x) = \sum_{n=0}^{\infty} \frac{x^n q^{n(n+1)}(1+q)(1+q^3)\ldots(1+q^{2n-1})}{(1-q^2)(1-q^4)\ldots(1-q^{2n})},$$

$$\mathscr{E}_2(x) = \sum_{n=0}^{\infty} \frac{x^n q^{n(n+1)}(1+q)(1+q^3)\ldots(1+q^{2n-1})(1+xq^{2n+1})}{(1-q^2)(1-q^4)\ldots(1-q^{2n})}.$$

8. Use Exercise 7 and your conjectures from Exercises 2 and 4 of Chapter 12 to conjecture, for $\mathscr{E}_2(1)$ and $\mathscr{E}_4(1)$, series-product identities similar to those of Theorem 14–6.

9. Use Exercise 9 of Section 13–2 to prove your conjectures in Exercise 8, thus proving also that $E_2(n) = p(W_2,n)$ and $E_4(n) = p(W_4,n)$.

10. Prove your conjecture for Exercise 5 of Chapter 12 using the technique employed in Theorem 14–6.

11. Let $e_6(l,n)$ denote the number of partitions of n in which no part is greater than l, in which the difference between any odd part and any other part that does not exceed that part is at least 3, and which does not contain 1. Prove that

$$e_6(l,n) = e_6(l-1,n) + e_6(l,n-2l) + e_6(l-2,n-2l+1).$$

12. Now let $\mathscr{E}_6(x) = \sum_{n=0}^{\infty} \sum_{l=0}^{\infty} e_6(l,n) q^n x^l$. Deduce from Exercise 11 that

$$\mathscr{E}_6(x)(1-x) = \mathscr{E}_6(xq^2)(1+x^2q^3).$$

13. Deduce from Exercise 12 that, for $|q| < 1$ and $|x| < 1$,

$$\mathscr{E}_6(x) = \prod_{n=0}^{\infty} \frac{(1+x^2q^{4n+3})}{(1-xq^{2n})}.$$

14. Prove that $\lim_{x\to 1^-} (1-x)\mathscr{E}_6(x) = \sum_{n=0}^{\infty} E_6(n) q^n$.

15. Use Exercises 13 and 14 to prove that $E_6(n) = p(W_6,n)$.

16. Let $\bar{E}_{12}(n)$ denote the number of partitions of n in which each part occurs at least twice. Use the concept of the conjugate partition to prove that $\bar{E}_{12}(n) = E_{12}(n)$.

17. Prove that, for $|q| < 1$,

$$\sum_{n=0}^{\infty} \bar{E}_{12}(n) q^n = \prod_{n=1}^{\infty} (1 + q^{2n} + q^{3n} + q^{4n} + \ldots).$$

18. Prove that, for $|q| < 1$,

$$\prod_{n=0}^{\infty} (1 + q^{2n} + q^{3n} + q^{4n} + \ldots) = \prod_{n=1}^{\infty} \frac{(1-q^n+q^{2n})}{(1-q^n)}$$

$$= \prod_{n=1}^{\infty} \frac{(1+q^{3n})}{(1-q^{2n})}.$$

19. Use Exercises 16, 17, and 18 to prove that $E_{12}(n) = p(W_{12}, n)$.

20. Let $U(n)$ denote the number of partitions of n in which no parts are congruent to 2 modulo 4, the difference between any two parts is at least 4, and no consecutive multiples of 4 appear. Let $Y(n)$ denote the number of partitions of n into distinct odd parts. Following the procedure used to prove Schur's Theorem (Theorem 14–8), prove that $U(n) = Y(n)$.

PART IV

GEOMETRIC NUMBER THEORY

Having studied both multiplicative and additive problems, we shall now use geometry to obtain information important to each area. The main relationship between number theory and geometry lies in analytic geometry, the association between ordered pairs of numbers (x,y) and points on a plane. For example, in the first quadrant, the number of lattice points (see Section 2–3 for the definition) lying on the hyperbola

$$xy = n$$

is clearly just $d(n)$, a function of importance in Part I. Similarly, the number of lattice points lying on the circle

$$x^2 + y^2 = n$$

is $r_2(n)$, the number of representations of n as a sum of two squares, a subject considered in Part III. Both of these geometric sets with their corresponding number-theoretic functions will be explored in Chapter 15.

CHAPTER 15

LATTICE POINTS

As you recall, a lattice point is a point in the *xy*-plane with integral coordinates. You have seen the importance of these points in the study of the linear Diophantine equation and in our proof of the Quadratic Reciprocity Law (Chapter 9). The advanced study of lattice points composes an entire branch of number theory, the Geometry of Numbers. We shall merely sample this subject by considering two elementary problems:

1. How large is $r_2(n)$, the number of ways of representing n as a sum of two squares?

2. How large is $d(n)$, the number of divisors of n?

We cannot obtain exact answers to these questions, but will have to settle for average values of $r_2(n)$ and $d(n)$; that is, for estimates of $\frac{1}{N}\sum_{n=0}^{N} r_2(n)$ and $\frac{1}{N}\sum_{n=1}^{N} d(n)$. Our estimates will be fairly accurate.

15–1 GAUSS'S CIRCLE PROBLEM

The problem of estimating $\frac{1}{N}\sum_{n=0}^{N} r_2(n)$ is called Gauss's Circle Problem. In Table 15–1, note that we count $a^2+b^2=n$ and $c^2+d^2=n$ as distinct representations of n if either $a \neq c$ or $b \neq d$, or both. Thus $r_2(1)=4$, since $1=1^2+0^2=0^2+1^2=0^2+(-1)^2=(-1)^2+0^2$.

Apparently, $r_2(n)$ assumes values irregularly, while the *average* value of $r_2(n)$ $\left(\frac{61}{19}=3.21 \text{ in Table 15–1}\right)$ behaves more regularly, tending to π as $n \to \infty$.

Theorem 15–1: $\lim_{N\to\infty} \frac{1}{N}\sum_{n=0}^{N} r_2(n) = \pi.$

TABLE 15–1: VALUES FOR $r_2(n)$ AND $\frac{1}{N}\sum_{n=0}^{N} r_2(n)$

n	$r_2(n)$	$\frac{1}{N}\sum_{n=0}^{N} r_2(n)$
0	1	–
1	4	5.00
2	4	4.50
3	0	3.00
4	4	3.25
5	8	4.20
6	0	3.50
7	0	3.00
8	4	3.13
9	4	3.22
10	8	3.70
11	0	3.36
12	0	3.08
13	8	3.46
14	0	3.21
15	0	3.00
16	4	3.06
17	8	3.35
18	4	3.39
19	0	3.21

PROOF: Let $\mathscr{C}(N)$ denote the set of all points in the xy-plane, either inside or on the circle whose equation is $x^2+y^2=N$. Since $r_2(n)$ is the number of lattice points on the circle $x^2+y^2=n$, we see that $\sum_{n=0}^{N} r_2(n)$ is equal to the number of lattice points in $\mathscr{C}(N)$. Figure 15–1 illustrates the case $N=4$.

With each lattice point Q, in or on $\mathscr{C}(N)$, let us associate a unit square of which Q is the upper left hand corner (for $Q=(1,1)$, the associated square is shaded in Figure 15–2). Let $P(N)$ denote the region comprising all these unit squares. The shaded area in Figure 15–3 is $P(4)$.

If $\mathscr{A}(R)$ denotes the area of the region R, then $\mathscr{A}(P(N))$ equals the number of lattice points in $\mathscr{C}(N)$, for each lattice point of $\mathscr{C}(N)$ contributes one unit square to the region $P(N)$. In Figure 15–3, we see that there are 13 lattice points in $\mathscr{C}(4)$, and that $\mathscr{A}(P(4))=13$. Hence, in general

$$\mathscr{A}(P(N)) = \sum_{n=0}^{N} r_2(n). \tag{15-1-1}$$

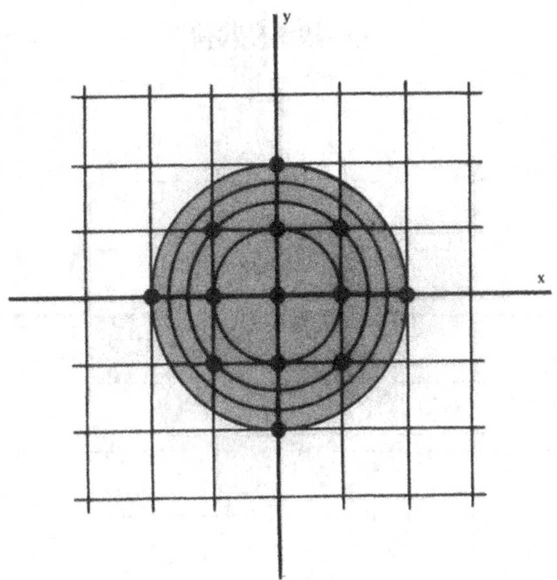

n	$r_2(n)$ (Number of Lattice Points on the Circle $x^2 + y^2 = n$)
0	1
1	4
2	4
3	0
4	4

Figure 15–1 The region $\mathscr{C}(4)$.

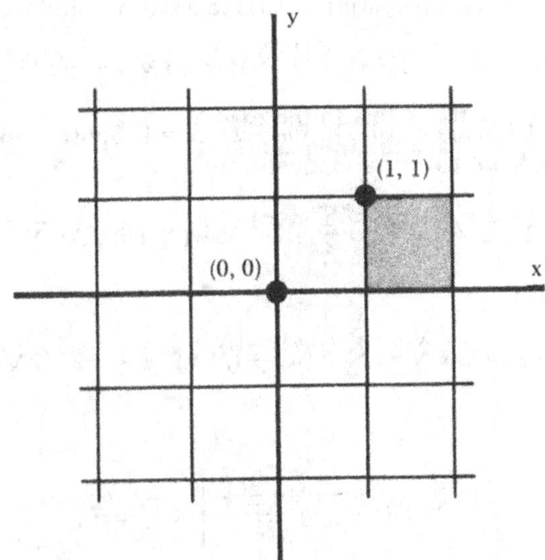

Figure 15–2 Unit square associated with (1,1).

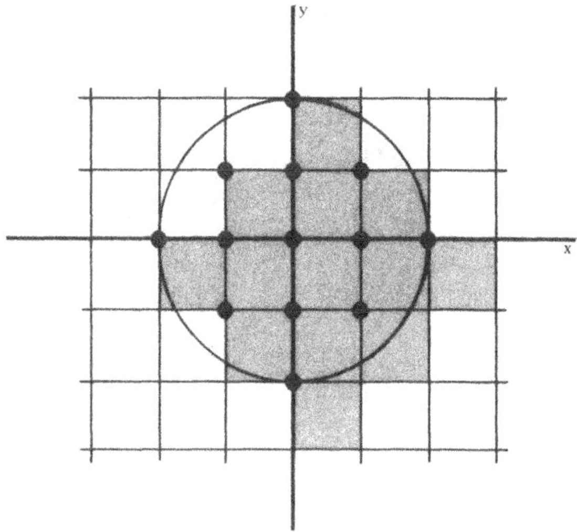

Figure 15–3 The region $P(4)$.

On the other hand, we see that any point on the boundary of $P(N)$ must be at most $\sqrt{2}$ units away from the boundary of $\mathscr{C}(N)$, for each unit square has a diagonal of length $\sqrt{2}$. Thus, $P(N)$ is contained in the circle centered on the origin of radius $\sqrt{N}+\sqrt{2}$, and, at the same time, it *contains* the circle centered on the origin of radius $\sqrt{N}-\sqrt{2}$. Hence, comparing the three areas, we find that

$$\pi(\sqrt{N}-\sqrt{2})^2 < \mathscr{A}(P(N)) < \pi(\sqrt{N}+\sqrt{2})^2. \qquad (15\text{-}1\text{-}2)$$

Figure 15–4 illustrates this in the case $N=4$. Substituting (15–1–1) into (15–1–2), we have the inequalities

$$\pi(N-2\sqrt{2}\sqrt{N}+2) < \sum_{N=0}^{N} r_2(n) < \pi(N+2\sqrt{2}\sqrt{N}+2);$$

that is,

$$-\pi 2\sqrt{2}\sqrt{N} < \sum_{n=0}^{N} r_2(n) - (N+2)\pi < \pi 2\sqrt{2}\sqrt{N}.$$

Thus,

$$\left|\frac{1}{N}\sum_{n=0}^{N} r_2(n) - \frac{(N+2)\pi}{N}\right| < \frac{\pi 2\sqrt{2}}{\sqrt{N}}, \qquad (15\text{-}1\text{-}3)$$

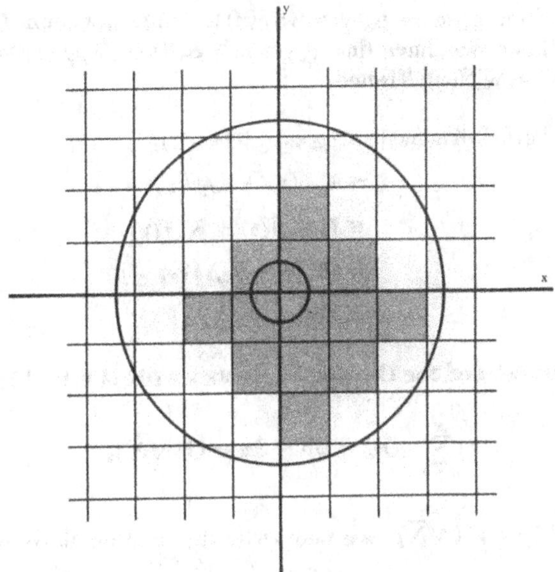

Figure 15-4 Circles containing, and contained in, $P(4)$.

and this implies that

$$\lim_{N\to\infty} \frac{1}{N} \sum_{n=0}^{N} r_2(N) = \lim_{N\to\infty} \frac{(N+2)\pi}{N} = \pi. \qquad \blacksquare$$

By means of some useful notation, we may modify the statement of Theorem 15-1.

DEFINITION 15-1: *We say that $f(N) = O(g(N))$ [read $f(N)$ is big "oh" of $g(N)$] if there exists a constant K such that*

$$|f(N)| \le Kg(N)$$

for all sufficiently large N.

While there are many important properties of the O-notation, we shall need only the following:

LEMMA: *If $h(x) = O(f(x))$, then*

$$O(h(x)) + O(f(x)) = O(f(x)).$$

PROOF: We have to prove that if $g_1(x) = O(h(x))$ and $g_2(x) =$

$O(f(x))$, then $g_1(x)+g_2(x)=O(f(x))$. This assertion is almost obvious, since we have that $|g_1(x)|\le K_1h(x)$, $|g_2(x)|\le K_2f(x)$, and $|h(x)|\le K_3f(x)$. Hence

$$\begin{aligned}|g_1(x)+g_2(x)| &\le |g_1(x)|+|g_2(x)|\\ &\le K_1h(x)+K_2f(x)\\ &\le K_1K_3f(x)+K_2f(x)\\ &=(K_1K_3+K_2)f(x)\\ &=Kf(x).\end{aligned}$$

■

Having defined the O-notation, we may write (15–1–3) as

$$\sum_{n=0}^{N} r_2(n)-\pi N-2\pi=O(\sqrt{N}),$$

and, since $2\pi=O(\sqrt{N})$, we may write the relation above as

$$\sum_{n=0}^{N} r_2(n)=\pi N+O(\sqrt{N}). \tag{15-1-4}$$

This equation asserts that $\sum_{n=0}^{N} r_2(n)$ approximately equals πN, with an error no larger than a constant multiple of $\sqrt{N}$.

Gauss's contribution to the Circle Problem ended with (15–1–4). Seventy-odd years elapsed before Sierpinski, in 1906, proved the far stronger result

$$\sum_{n=0}^{N} r_2(n)=\pi N+O(N^{1/3}). \tag{15-1-5}$$

Later mathematicians have proved this equation to be correct even when 1/3 is replaced by a somewhat smaller number, for example 27/82. It is known that (15–1–5) is false when the exponent is 1/4, but the smallest number that can replace 1/3 has not been determined.

EXERCISES

1. Prove that if $r_3(n)$ denotes the number of representations of n as a sum of three squares, then

$$\sum_{n=0}^{N} r_3(n)=\frac{4}{3}\pi N^{3/2}+O(N).$$

[Hint: Count the lattice points in a sphere of radius $\sqrt{N}$ centered on the origin.]

2. Prove that the number of lattice points in the region

$$|y| + x^2 \leq N$$

is $\frac{8}{3} N^{3/2} + O(N)$.

15–2 DIRICHLET'S DIVISOR PROBLEM

Our object here is to find an average value for the number $d(n)$ of divisors of n.

THEOREM 15–2: *There exists a constant c such that*

$$\sum_{n=0}^{N} d(n) = N \log N + cN + O(\sqrt{N}).$$

PROOF: The function $d(n)$ counts the number of lattice points of the form (x,y) where $x > 0$, $y > 0$, and $xy = n$. Thus, $d(n)$ counts the number of lattice points lying on the hyperbola $xy = n$ in the first quadrant. Figure 15–5 illustrates the hyperbolas for $n = 1, 2, 3, 4, 5,$ and 6. We see that $\sum_{n=1}^{N} d(n)$ is the number of lattice points with positive coordinates lying under or on the hyperbola $xy = N$.

Let us now carefully examine Figure 15–6. The shaded square region clearly contains $[\sqrt{N}]^2$ lattice points*; two identically shaped regions, R_1 above the square and R_2 to the right, contain the remaining lattice points. Now, for $1 \leq n \leq \sqrt{N}$, there are $\left(\left[\frac{N}{n}\right] - [\sqrt{N}]\right)$ lattice points in R_1 lying on the vertical line $x = n$. Since the total number of lattice points in R_2 must equal the number in R_1, we have the equation

$$\sum_{n=1}^{N} d(n) = 2 \sum_{n=1}^{[\sqrt{N}]} \left(\left[\frac{N}{n}\right] - [\sqrt{N}]\right) + [\sqrt{N}]^2. \qquad (15\text{-}2\text{-}1)$$

Now let us define $\{x\}$, the fractional part of x, by $\{x\} = x - [x]$.

---

* $[\sqrt{N}]$ denotes the largest integer not exceeding $\sqrt{N}$.

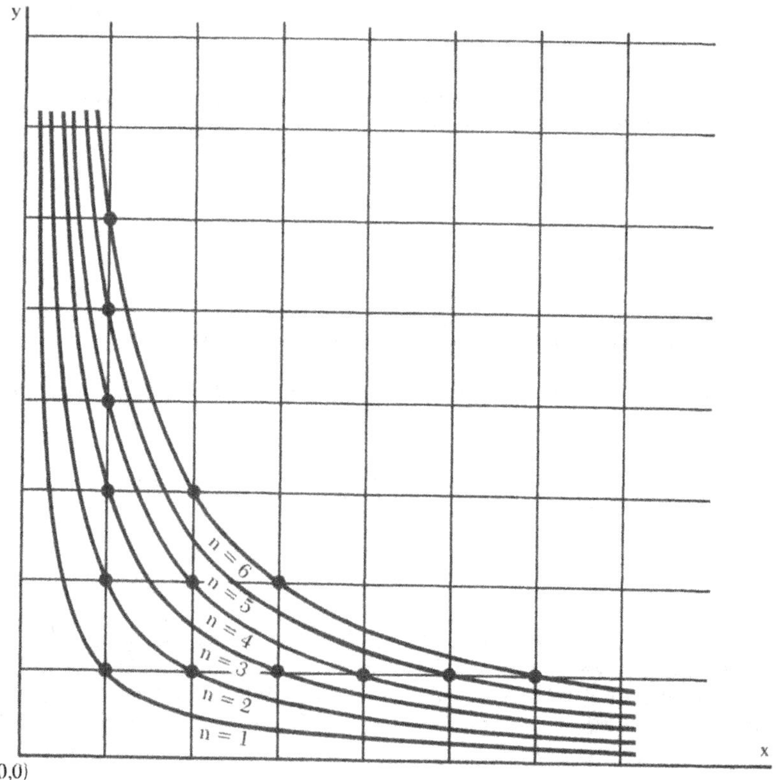

Figure 15–5

Clearly $0 \leq \{x\} < 1$. Hence,

$$\begin{aligned}
\sum_{n=1}^{N} d(n) &= 2\sum_{n=1}^{[\sqrt{N}]} \left(\left[\frac{N}{n}\right] - [\sqrt{N}]\right) + [\sqrt{N}]^2 \\
&= 2\sum_{n=1}^{[\sqrt{N}]} \left[\frac{N}{n}\right] - 2[\sqrt{N}]^2 + [\sqrt{N}]^2 \\
&= 2\sum_{n=1}^{[\sqrt{N}]} \left[\frac{N}{n}\right] - [\sqrt{N}]^2 \\
&= 2\sum_{n=1}^{[\sqrt{N}]} \left(\frac{N}{n} - \left\{\frac{N}{n}\right\}\right) - (\sqrt{N} - \{\sqrt{N}\})^2 \\
&= 2N\sum_{n=1}^{[\sqrt{N}]} \frac{1}{n} + O(\sqrt{N}) - N + O(\sqrt{N})
\end{aligned}$$

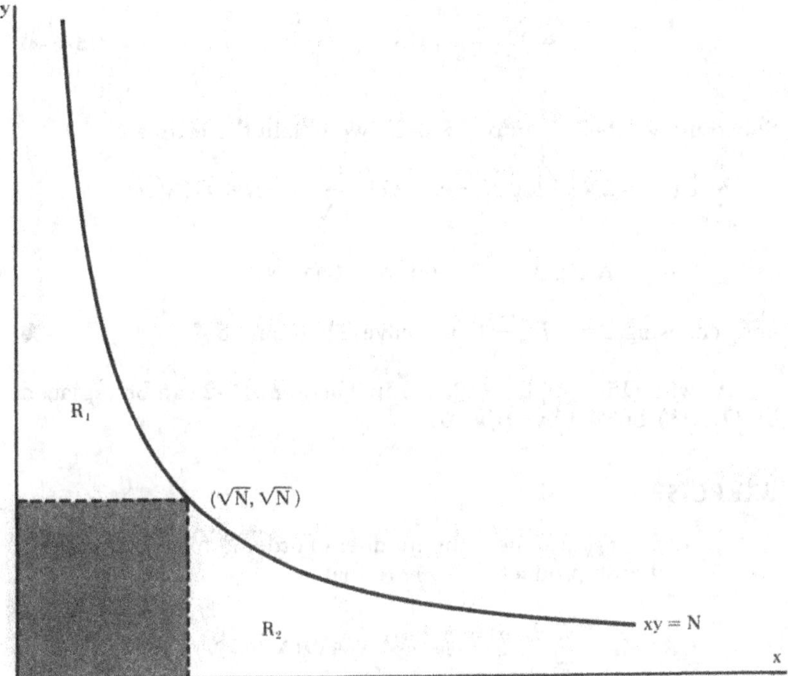

Figure 15-6

$$= 2N \sum_{n=1}^{[\sqrt{N}]} \frac{1}{n} - N + O(\sqrt{N}). \tag{15-2-2}$$

To finish the proof of Theorem 15-2, we must use the integral test from calculus (see Appendix D). This test asserts that, if $f(t)$ is a positive decreasing function of t such that $\lim_{t\to\infty} f(t) = 0$, then there exists a constant k such that, for each $M > 1$,

$$\int_1^M f(t)\,dt = \sum_{n=1}^{M} f(n) + k + O(f(M)). \tag{15-2-3}$$

Now, if $f(t) = \frac{1}{t}$ and $M = \sqrt{N}$, then

$$\frac{1}{2}\log N = \log \sqrt{N}$$

$$= \int_1^{\sqrt{N}} \frac{1}{t}\,dt$$

$$= \sum_{n=1}^{[\sqrt{N}]} \frac{1}{n} + c_0 + O\left(\frac{1}{\sqrt{N}}\right). \tag{15-2-4}$$

Substituting (15–2–4) into (15–2–2), we obtain the estimate

$$\sum_{n=1}^{N} d(n) = 2N\left(\frac{1}{2}\log N - c_0 - O\left(\frac{1}{\sqrt{N}}\right)\right) - N + O(\sqrt{N})$$

$$= N \log N - (2c_0 + 1)N + O(\sqrt{N}),$$

and, choosing $c = -2c_0 - 1$, we have Theorem 15–2. ■

As with (15–1–4), the $O(\sqrt{N})$ in Theorem 15–2 can be replaced by $O(N^{1/3})$ but not by $O(N^{1/4})$.

EXERCISE

1. Let $d_3(n)$ denote the number of ordered triples of integers whose product is n. Prove that

$$\sum_{n=1}^{N} d_3(n) = N \log^2 N + O(N \log N).$$

[Hint: Count the lattice points with positive coordinates on or under the surface defined by $xyz = N$.]

APPENDICES

APPENDIX A:

A PROOF THAT $\lim_{n\to\infty} p(n)^{1/n} = 1$

Lemma 1: *The relation $p_m(n) \leq (n+1)^m$ holds for each integer $m > 0$.*

Proof: If, in the expression $a_1 + a_2 + \ldots + a_m$, each a_i takes all integral values in the intervals $[0, n]$, we obtain all decompositions of n into at most m parts, as well as many decompositions of other numbers. Since there are $(n+1)^m$ possible decompositions, the lemma follows. ■

Lemma 2: *The relation $p(n) \leq p(n-1) + p_m(n) + p(n-m)$ holds for each integer $m > 0$.*

Proof: Separate the partitions of n into 3 classes: the partitions of the first class contain 1 as a summand; those in the second class contain no 1's and have at most m parts; and those in the third class contain no 1's and have more than m parts.

Deletion of a 1 from each partition of the first class leaves exactly $p(n-1)$ partitions. The second class clearly contains at most $p_m(n)$ elements. In the third class, subtract 1 from the smallest m summands of each partition; this establishes a one-to-one correspondence between the elements of the third class and a subset of the partitions of $n - m$. Hence, the third class has at most $p(n-m)$ elements. Consequently, $p(n) \leq p(n-1) + p_m(n) + p(n-m)$, as was to be proved. ■

Theorem A-1: $\lim_{n\to\infty} p(n)^{1/n} = 1.$

Proof: It is sufficient to show that, for each $\epsilon > 0$,

$$p(n) < K(1+\epsilon)^n \tag{A-1}$$

if n is sufficiently large; this will imply directly that

$$1 \leq p(n)^{1/n} < K^{1/n}(1+\epsilon) \to 1+\epsilon \quad \text{as} \quad n \to \infty.$$

Choose m sufficiently large so that $(1+\epsilon)^{m-1} > 2/\epsilon$. Next, choose n_0 so large that, for $n > n_0$,

$$p_m(n) < \frac{\epsilon}{2}(1+\epsilon)^{n-1};$$

this is possible, since $p_m(n) \leq (n+1)^m$ by Lemma 1, and since

$$\lim_{n\to\infty} \frac{(n+1)^m}{(1+\epsilon)^{n-1}} = 0,$$

by an m-fold application of L'Hospital's Rule.

Now let

$$K = \left(\max_{0 \leq n \leq n_0} \frac{p(n)}{(1+\epsilon)^n}\right) + 1.$$

Then $p(n) < K(1+\epsilon)^n$ for all $n \leq n_0$.

Assume that $p(n) < K(1+\epsilon)^n$ for all $n < N$, where $N > n_0$. Then

$$\begin{aligned} p(N) &\leq p(N-1) + p_m(N) + p(N-m) \\ &< K(1+\epsilon)^{N-1} + \frac{\epsilon}{2}(1+\epsilon)^{N-1} + K(1+\epsilon)^{N-m} \\ &< K(1+\epsilon)^{N-1}\left(1 + \frac{\epsilon}{2} + \frac{1}{(1+\epsilon)^{m-1}}\right) \\ &< K(1+\epsilon)^{N-1}\left(1 + \frac{\epsilon}{2} + \frac{\epsilon}{2}\right) = K(1+\epsilon)^N. \end{aligned}$$

Hence, by mathematical induction, (A–1) holds for all sufficiently large n. ∎

APPENDIX B:

INFINITE SERIES AND PRODUCTS

Convergence and Rearrangement of Series and Products

Recall the ratio test from calculus: The series $\sum_{n=0}^{\infty} a_n$ is absolutely convergent if $\lim_{n\to\infty} \left|\frac{a_{n+1}}{a_n}\right| < 1$.

This test is adequate to establish the convergence of the series in (13–2–1), (13–2–2), and all similar series appearing in Chapters 13 and 14.

For example, in (13–2–1), $\sum_{n=0}^{\infty} \frac{q^{\frac{1}{2}n(n-1)} z^n}{(1-q)\ldots(1-q^n)}$ is absolutely convergent for $|q| < 1$, since

$$\lim_{n\to\infty} \left|\frac{a_{n+1}}{a_n}\right| = \lim_{n\to\infty} \left| \frac{q^{\frac{1}{2}n(n+1)} z^{n+1}}{(1-q)\ldots(1-q^{n+1})} \Big/ \frac{q^{\frac{1}{2}n(n-1)} z^n}{(1-q)\ldots(1-q^n)} \right|$$

$$= \lim_{n\to\infty} \left| \frac{q^n z}{(1-q^{n+1})} \right| = 0, \text{ if } |q| < 1.$$

Similar results hold for the other series.

We now exhibit several relationships between infinite series and infinite products; they will establish the convergence of products in Chapters 13 and 14.

THEOREM B–1: *If $a_n \geq 0$, then $\prod_{n=1}^{\infty} (1+a_n)$ and $\sum_{n=1}^{\infty} a_n$ are both convergent or both divergent.*

PROOF: Since the function $g(x) = e^x - x - 1$ has an absolute minimum at $x = 0$ and since $g(0) = 0$, we see that $1 + x \leq e^x$ for all x. Hence,

$$1 + a_1 + a_2 + \ldots + a_N \leq \prod_{n=1}^{N} (1+a_n) \leq e^{a_1+a_2+\ldots+a_N}. \qquad \text{(B-1)}$$

Consequently, if either the sequence of partial sums or the sequence of partial products converges, then the other is bounded and so must converge because both are nondecreasing. Finally, $\prod_{n=1}^{\infty} (1 + a_n) \neq 0$ since each partial product is at least 1. ■

THEOREM B–2: *If* $1 > a_n \geq 0$, *then* $\prod_{n=1}^{\infty} (1 - a_n)$ *and* $\sum_{n=1}^{\infty} a_n$ *are both convergent or both divergent.*

PROOF: Suppose $\sum_{n=1}^{\infty} a_n$ converges; then there exists an N such that $\sum_{n=N}^{\infty} a_n < \frac{1}{2}$. Now

$$\begin{aligned}(1 - a_N)(1 - a_{N+1}) &= 1 - a_N - a_{N+1} + a_N a_{N+1} \\ &\geq 1 - a_N - a_{N+1}\end{aligned}$$

and

$$\begin{aligned}(1 - a_N)(1 - a_{N+1})(1 - a_{N+2}) &\geq (1 - a_N - a_{N+1})(1 - a_{N+2}) \\ &\geq 1 - a_N - a_{N+1} - a_{N+2}.\end{aligned}$$

We may proceed by mathematical induction to show that, for $m \geq N$,

$$(1 - a_N)(1 - a_{N+1}) \ldots (1 - a_m) \geq 1 - a_N - a_{N+1} \ldots - a_m.$$

Consequently, if $p_m = \prod_{n=1}^{m} (1 - a_m)$, then for $m \geq N$

$$\begin{aligned}\frac{p_m}{p_{N-1}} &= (1 - a_N)(1 - a_{N+1}) \ldots (1 - a_m) \\ &\geq 1 - a_N - a_{N+1} \ldots - a_m > 1 - \frac{1}{2} = \frac{1}{2}.\end{aligned}$$

Now $p_m > 0$ for all m, since p_m is a product of positive numbers. Furthermore,

$$p_m - p_{m+1} = p_m(1 - (1 - a_{m+1})) = a_{m+1} p_m \geq 0;$$

thus the p_m form a decreasing sequence. Also, $p_m > \frac{1}{2} p_{N-1} > 0$. Hence, the sequence is bounded below by a positive number. Since every decreasing sequence bounded below by a positive number tends to a positive limit, we see that $\lim_{m\to\infty} p_m$ exists and is not zero; that is, $\prod_{n=1}^{\infty}(1-a_n)$ converges.

Alternately, begin with the assumption that $\prod_{n=1}^{\infty}(1-a_n)$ converges. Then, since $1-x \le e^{-x}$ for all x (this is the inequality used in Theorem B–1, except that $-x$ replaces x), we have the inequalities

$$0 < c < (1-a_1)(1-a_2)\ldots(1-a_m) \le e^{-a_1-a_2-\ldots-a_m},$$

where $c < 1$. Thus,

$$\log c < -a_1 - a_2 - \ldots - a_m;$$

that is,

$$a_1 + a_2 + \ldots + a_m < -\log c.$$

Hence, $\left\{\sum_{n=1}^{m} a_n\right\}_{m=1}^{\infty}$ forms a bounded increasing sequence, and so $\sum_{n=1}^{\infty} a_n$ converges. ■

The next theorem justifies the manipulation of the products in Chapters 13 and 14. First we define absolute convergence.

DEFINITION B–1: *The infinite product* $\prod_{n=1}^{\infty}(1+a_n)$ *is said to converge absolutely if* $\prod_{n=1}^{\infty}(1+|a_n|)$ *converges.*

THEOREM B–3: *If* $|a_n| < 1$, *and* $\prod_{n=1}^{\infty}(1+a_n)$ *converges absolutely, then* $\prod_{n=1}^{\infty}(1+a_n)$ *converges.*

PROOF: Let $P_N = \prod_{n=1}^{N}(1+|a_n|)$ and $p_N = \prod_{n=1}^{N}(1+a_n)$. Then

$$\begin{aligned}|p_N - p_{N-1}| &= |(1+a_1)(1+a_2)\ldots(1+a_{N-1})a_N|\\ &\le (1+|a_1|)(1+|a_2|)\ldots(1+|a_{N-1}|)\,|a_N|\\ &= P_N - P_{N-1}.\end{aligned}$$

Consequently, if $R > S$,

$$\begin{aligned} |p_R - p_S| &= |p_R - p_{R-1} + p_{R-1} - p_{R-2} + \ldots + p_{S+2} - p_{S+1} + p_{S+1} - p_S| \\ &\leq |p_R - p_{R-1}| + |p_{R-1} - p_{R-2}| + \ldots + |p_{S+2} - p_{S+1}| \\ &\quad + |p_{S+1} - p_S| \\ &\leq P_R - P_{R-1} + P_{R-1} - P_{R-2} + \ldots + P_{S+2} - P_{S+1} + P_{S+1} - P_S \\ &= P_R - P_S. \end{aligned}$$

Now we know that $\lim_{N\to\infty} P_N$ exists, so that $\{P_N\}_{N=1}^{\infty}$ is a Cauchy sequence. (Recall from calculus that $\{\alpha_n\}_{n=1}^{\infty}$ is a Cauchy sequence, provided for each $\epsilon > 0$, there exists an M such that $|\alpha_R - \alpha_S| < \epsilon$ whenever $R \geq S \geq M$. Recall also that a sequence converges to a limit if and only if it is a Cauchy sequence.) Since $\{P_N\}_{N=1}^{\infty}$ is a Cauchy sequence, the inequality $|p_R - p_S| \leq P_R - P_S$ implies that $\{p_N\}_{N=1}^{\infty}$ is also a Cauchy sequence. Consequently, $\lim_{N\to\infty} p_N$ exists.

We shall now show that $\lim_{N\to\infty} p_N \neq 0$. Clearly,

$$\begin{aligned} |p_N| &= |(1 + a_1)(1 + a_2) \ldots (1 + a_N)| \\ &\geq (1 - |a_1|)(1 - |a_2|) \ldots (1 - |a_N|) > c > 0 \end{aligned}$$

since, by Theorem B–2, $\prod_{n=1}^{\infty} (1 - |a_n|)$ is convergent. Hence $|\lim_{N\to\infty} p_N| \geq c$. ■

We now apply Theorem B–3 to some of the infinite products appearing in Chapters 13 and 14.

A product such as $\prod_{n=1}^{\infty} (1 + zq^{5n+4})$ is absolutely convergent for $|q| < 1$ since

$$\sum_{n=1}^{\infty} |zq^{5n+4}| = |z| \frac{|q^4|}{1 - |q^5|}.$$

A product such as $\prod_{n=1}^{\infty} \frac{1}{1 - zq^n}$ is absolutely convergent for $|q| < 1$ and $|z| < |q^{-1}|$, since

$$\prod_{n=1}^{\infty} \frac{1}{1 - zq^n} = \prod_{n=1}^{\infty} \left(1 + \frac{zq^n}{1 - zq^n}\right),$$

and since $\sum_{n=1}^{\infty} \frac{zq^n}{1-zq^n}$ is absolutely convergent by comparison with $\frac{1}{1-|zq|} \sum_{n=1}^{\infty} |zq|^n$.

If a_n is real, $|a_n| < 1$, and $\sum_{n=1}^{\infty} |a_n|$ converges, then $\prod_{n=1}^{\infty} (1+a_n) = \exp\left(\sum_{n=1}^{\infty} \log(1+a_n)\right)$. Now, $\sum_{n=1}^{\infty} \log(1+a_n)$ converges absolutely, since $\sum_{n=1}^{\infty} |a_n|$ converges. We can therefore justify rearranging the terms of an absolutely convergent infinite *product* by the argument that rearrangement of an absolutely convergent infinite *series* does not alter the value of the sum.

Maclaurin Series Expansion of Infinite Products

In order to justify the use of Maclaurin series expansions for infinite products in Theorem 13–7 or in Theorem 14–3, we could introduce the subject of uniformly convergent infinite products. However, there is a simpler procedure. Let us execute, without special assumptions, a "reverse proof" of (13–2–1).

Let $F(z) = \sum_{n=0}^{\infty} \frac{q^{\frac{1}{2}n(n-1)} z^n}{(1-q)\ldots(1-q^n)}$, where, by convention, the first term is equal to 1. Then

$$(1+z)F(zq) = \sum_{n=0}^{\infty} \frac{q^{\frac{1}{2}n(n+1)} z^n}{(1-q)\ldots(1-q^n)} + \sum_{n=0}^{\infty} \frac{q^{\frac{1}{2}n(n+1)} z^{n+1}}{(1-q)\ldots(1-q^n)}$$

$$= \sum_{n=0}^{\infty} \frac{q^{\frac{1}{2}n(n+1)} z^n}{(1-q)\ldots(1-q^n)} + \sum_{n=1}^{\infty} \frac{q^{\frac{1}{2}n(n-1)} z^n}{(1-q)\ldots(1-q^{n-1})}$$

$$= 1 + \sum_{n=1}^{\infty} \frac{q^{\frac{1}{2}n(n-1)} z^n}{(1-q)\ldots(1-q^{n-1})} \left(\frac{q^n}{1-q^n} + 1\right)$$

$$= 1 + \sum_{n=1}^{\infty} \frac{q^{\frac{1}{2}n(n-1)} z^n}{(1-q)\ldots(1-q^n)}$$

$$= F(z).$$

Thus,

$$\begin{aligned} F(z) &= (1+z)F(zq) \\ &= (1+z)(1+zq)F(zq^2) \\ &\ldots \\ &= (1+z)(1+zq)\ \ldots\ (1+zq^N)F(zq^{N+1}). \end{aligned}$$

This series for $F(z)$ converges uniformly for $|z| < M$ and $|q| < 1$, by the ratio test; therefore, $F(z)$ is continuous at $z = 0$. Hence, since $F(0) = 1$,

$$\begin{aligned} F(z) &= \lim_{N\to\infty} \frac{F(z)}{F(zq^{N+1})} \\ &= \lim_{N\to\infty} \prod_{n=0}^{N} (1+zq^n) \\ &= \prod_{n=0}^{\infty} (1+zq^n). \end{aligned}$$

As you can see, we've made no assumption about uniform convergence of infinite products; all we require is uniform convergence of infinite series.

APPENDIX C:

DOUBLE SERIES

Chapters 13 and 14 contain expressions of the form

$$\sum_{n=0}^{\infty} \sum_{m=0}^{\infty} a_{m,n};$$

in the proof of Theorem 13–8, we use the rearrangement

$$\sum_{n=0}^{\infty} \sum_{m=0}^{\infty} a_{m,n} = \sum_{m=0}^{\infty} \sum_{n=0}^{\infty} a_{m,n},$$

which is permissible only under certain conditions.

Our use of double series is justified by the theorems to follow.

In treating *ordinary* infinite series, we define $\sum_{n=0}^{\infty} \alpha_n$ by the formula

$$\sum_{n=0}^{\infty} \alpha_n = \lim_{N\to\infty} \left(\sum_{n=0}^{N} \alpha_n\right).$$

There are, however, many possible definitions for the sum of a *double* series, $\sum_{m=0}^{\infty} \sum_{n=0}^{\infty} a_{m,n}$. We may consider summation by rows or by columns; for example

$$\lim_{M\to\infty} \sum_{m=0}^{M} \sum_{n=0}^{\infty} a_{m,n}, \text{ or } \lim_{N\to\infty} \sum_{n=0}^{N} \sum_{m=0}^{\infty} a_{m,n}.$$

In the first case, if the limit is finite, we say that the double series is *convergent by rows;* in the second case, that it is *convergent by columns.*

Theorem C–1: *If $a_{m,n} \geq 0$ for all $m \geq 0$ and all $n \geq 0$, then*

$$\lim_{M\to\infty} \sum_{m=0}^{M} \sum_{n=0}^{\infty} a_{m,n} = \lim_{N\to\infty} \sum_{n=0}^{N} \sum_{m=0}^{\infty} a_{m,n},$$

(if either limit is $+\infty$, so is the other).

PROOF: Suppose that

$$\lim_{M\to\infty} \sum_{m=0}^{M} \sum_{n=0}^{\infty} a_{m,n} = S < \infty.$$

Let $A_m = \sum_{n=0}^{\infty} a_{m,n}$; clearly this series converges for each m. Thus

$$\sum_{m=0}^{\infty} A_m = S.$$

Now, $a_{m,n} \leq A_m$; hence, by the comparison test,

$$\sum_{m=0}^{\infty} a_{m,n} \leq S.$$

Let $B_n = \sum_{m=0}^{\infty} a_{m,n}$; then

$$\sum_{n=0}^{N} B_n = \sum_{n=0}^{N} \sum_{m=0}^{\infty} a_{m,n} = \sum_{m=0}^{\infty} \sum_{n=0}^{N} a_{m,n} \leq \sum_{m=0}^{\infty} A_m = S.$$

Consequently, $\lim_{N\to\infty} \sum_{n=0}^{N} B_n$ exists; let us call this limit S'. Clearly, $S' \leq S$.

Now we may reverse the argument, and start with

$$\lim_{N\to\infty} \sum_{n=0}^{N} \sum_{m=0}^{\infty} a_{m,n} = S'.$$

We can then deduce that $S \leq S'$. Hence $S = S'$, and our theorem is established. ∎

COROLLARY: *If $0 \leq b_{m,n} \leq a_{m,n}$, and if $\sum\sum a_{m,n}$ is convergent by either rows or columns, then $\sum\sum b_{m,n}$ is convergent by both rows and columns, the two resulting sums being identical.*

PROOF: $\sum_{m=0}^{M} \sum_{n=0}^{\infty} b_{m,n}$ and $\sum_{n=0}^{N} \sum_{m=0}^{\infty} b_{m,n}$ are both increasing sequences bounded above by $\sum\sum a_{m,n}$; consequently each converges. Theorem C-1 implies that the sums are identical, and this allows us to make the following convention. ∎

DEFINITION C–1: *In the light of Theorem C–1, if $a_{m,n} \geq 0$, and $\sum\sum a_{m,n}$ converges by either rows or columns, we simply say that it* **converges.**

DEFINITION C–2: *If $\sum\sum |a_{m,n}|$ converges, we shall say that* $\sum\sum a_{m,n}$ **converges absolutely.**

THEOREM C–2: *If $\sum\sum a_{m,n}$ converges absolutely, then*

$$\lim_{M\to\infty} \sum_{m=0}^{M} \sum_{n=0}^{\infty} a_{m,n} = \lim_{N\to\infty} \sum_{n=0}^{N} \sum_{m=0}^{\infty} a_{m,n}$$

and both limits are finite.

PROOF: We define

$$\alpha_{m,n} = \begin{cases} a_{m,n} & \text{if } a_{m,n} \geq 0, \\ 0 & \text{otherwise;} \end{cases}$$

$$\beta_{m,n} = \begin{cases} -a_{m,n} & \text{if } a_{m,n} \leq 0, \\ 0 & \text{otherwise.} \end{cases}$$

Clearly, by comparison with $\sum\sum |a_{m,n}|$, we see that $\sum\sum \alpha_{m,n}$ and $\sum\sum \beta_{m,n}$ converge (by both rows and columns). Hence, $\sum\sum (\alpha_{m,n} - \beta_{m,n}) = \sum\sum a_{m,n}$ converges by both rows and columns. ■

As in the corollary to Theorem C–1, if $|b_{m,n}| \leq a_{m,n}$, and if $\sum\sum a_{m,n}$ converges, then $\sum\sum b_{m,n}$ converges.

We are now prepared to establish a result that justifies the double series expansions in Chapter 14.

THEOREM C–3: *Suppose that*

$$A_m(q) = \sum_{n=0}^{\infty} a_{m,n} q^n \text{ (where } a_{m,n} \geq 0)$$

converges absolutely for $|q| < 1$. Suppose also that $a_{m,n} = 0$ for $m > n$, and that $a_{m,m+n} \leq Kc_n$, where K is an absolute positive constant and $\sum_{n=0}^{\infty} c_n z^n$ converges absolutely for $|z| < 1$. Then

$$\sum_{m=0}^{\infty}\sum_{n=0}^{\infty} a_{m,n}x^m q^n$$

converges absolutely for $|q| < 1$ *and* $|x| < |q|^{-1}$.

PROOF:

$$\sum_{m=0}^{\infty}\left(\sum_{n=0}^{\infty} |a_{m,n}x^m q^n|\right) = \sum_{m=0}^{\infty}\left(\sum_{n=m}^{\infty} |a_{m,n}x^m q^n|\right)$$

$$= \sum_{m=0}^{\infty}\left(\sum_{n=0}^{\infty} |a_{m,n+m}x^m q^{n+m}|\right)$$

$$\leq \sum_{m=0}^{\infty}\sum_{n=0}^{\infty} Kc_n|xq|^m|q|^n = \frac{K}{1-|xq|}\sum_{n=0}^{\infty} c_n|q|^n,$$

where the convergence of the last series requires that $|q| < 1$ and $|x| < |q|^{-1}$. ∎

Let us now apply Theorem C–3 to the expansions (14–3–4) and (14–3–5).

The partition function $\delta_i(m,n)$ takes the value 0 if $m > n$, since a partition of n into m positive parts implies that $n \geq m$; furthermore,

$$\begin{aligned}\delta_i(m,n+m) &\leq \delta_1(m,n+m)\\ &= \delta_1(m,n) + \delta_0(m-1,n) \qquad \text{(by (14–3–1) and (14–3–2))}\\ &\leq 2p(n).\end{aligned}$$

Hence, Theorem C–3 applies to (14–3–4) and (14–3–5) with $K = 2$ and $c_n = p(n)$.

A similar argument establishes the expansion used in the proof of Schur's theorem:

$$H_i(x;q) = \sum_{M=0}^{\infty}\sum_{N=0}^{\infty} \Delta_i(M,N)x^M q^N.$$

Here, $\Delta_i(M,N) \leq \delta_1(M,N)$, and the comparison test applies directly.

Finally, we show how to establish (14–3–15). First, applying the ratio test to (14–3–9), we find that

$$\lim_{n\to\infty}\left|\frac{x^2q^{5n+4-i}(1-x^{i+1}q^{(2n+3)(i+1)})}{(1-q^{n+1})(1-x^{i+1}q^{(2n+1)(i+1)})(1-xq^n)}\right| = 0,$$

since $|q| < 1$. Thus, if $|q| < 1$ (and $|x| < |q|^{-1}$, so that the denominator of the series for $f_i(x;q)$ is always defined), then the series for $f_i(x;q)$ converges. In fact, it converges uniformly for $|q| \leq 1-\epsilon$ and $|x| \leq |q|^{-1}-\epsilon$; therefore, $f_i(x;q)$ is an analytic function of x for $|q| < 1$ and $|x| < |q|^{-1}$. We can now expand $f_i(x;q)$ for $i=1$ and 0 (as in Section 14–4) using only (14–3–13) and (14–3–12). Thus, as in Section 14–4,

$$f_1(x;q) = \sum_{m=0}^{\infty} \frac{q^{m^2}x^m}{(1-q)\dots(1-q^m)},$$

$$f_0(x;q) = \sum_{m=0}^{\infty} \frac{q^{m^2+m}x^m}{(1-q)\dots(1-q^m)}.$$

Hence, the coefficient $c_i(m,n)$ of $x^m q^n$ in (14–3–15) is $\pi_m(n-m^2)$ when $i=1$ and $\pi_m(n-m^2-m)$ when $i=0$. In either case $c_i(m,n) \leq p(n)$; also $c_i(m,n)=0$ if $n < m$.

APPENDIX D:

THE INTEGRAL TEST

THEOREM D–1: *If $f(t)$ is a positive decreasing function of t such that $\lim_{t\to\infty} f(t) = 0$, then there exists a constant k such that, for each $M > 1$,*

$$\int_1^M f(t)\,dt = \sum_{n=1}^{M} f(n) + k + O(f(M)).$$

PROOF: Let $N = [M]$. Then

$$\int_1^N f(t)\,dt - \sum_{n=1}^{N} f(n) = \sum_{n=2}^{N}\left(\int_{n-1}^{n} f(t)\,dt - f(n)\right) - f(1)$$

$$= \sum_{n=2}^{N} a_n - f(1),$$

where

$$a_n = \int_{n-1}^{n} (f(t) - f(n))\,dt \geq 0.$$

Now

$$0 \leq \sum_{n=R+1}^{S} a_n \leq \sum_{n=R+1}^{S} (f(n-1) - f(n)) = f(R) - f(S).$$

Thus $\sum_{n=2}^{N} a_n$ is a bounded series of nonnegative terms, and therefore it converges. Furthermore, the string of relations in the preceding paragraph shows that $\sum_{n=M+1}^{\infty} a_n = O(f(M))$. Thus

$$\int_1^M f(t)\,dt = \sum_{n=1}^{M} f(n) + \left(\sum_{n=2}^{\infty} a_n - f(1)\right) + O(f(M))$$

$$= \sum_{n=1}^{M} f(n) + k + O(f(M)).$$ ■

NOTES

(See Bibliography for full details on all references)

Chapter 1

The proof of the basis representation theorem (Theorem 1–3) is by Andrews (1969).

Chapter 2

Much of this material is historically developed in Dickson (1952b), Chapter 2.

Chapter 3

Riordan (1958) presents an extensive introduction to combinatorial analysis, including many applications of generating functions.

The proof of Fermat's little theorem (Theorem 3–4) was originally given by Golomb (1956).

Carmichael (1959), Part I, Chapter 4, gives our proof of Wilson's theorem.

A thorough account of the relation between elementary number theory and computing is given by Knuth (1969), Chapter 4.

The history of the material in this chapter may be found in Dickson (1952a), Chapter 3.

Chapter 4

The applications of congruences to card shuffling are discussed by Uspensky and Heaslet (1939) and Johnson (1956). Dickson (1952b), Chapter 2, contains a historical development of congruences.

Chapter 5

See the notes on Chapter 3 for references to the theorems of Fermat and Wilson.

The background of the Chinese Remainder Theorem is given in Dickson (1952b), Chapter 2.

Lagrange's theorem is discussed in Dickson (1952a), Chapter 8.

Chapter 6

Ths history of the arithmetical functions $\phi(n)$, $d(n)$, $\sigma(n)$, and $\mu(n)$ is found in Chapters 5, 2, 2, and 19, respectively, of Dickson (1952a).

Chapter 7

The theory of primitive roots is developed historically in Dickson (1952a), Chapter 7.

Chapter 8

The proof that $\lim_{x\to\infty} \pi(x)/x=0$ (Theorem 8–4) simplifies a version by Mamangakis (1962).

Dickson (1952b), Chapter 18, gives the history of primes prior to 1920; more recent developments are discussed by Grosswald (1966), Chapters 7 and 8.

Chapter 9

Our proof of the Quadratic Reciprocity Law was originally discovered by D. H. Lehmer (1957). Quadratic residues are discussed by Dickson (1952a), Chapter 8.

Chapter 10

Sums of the form $\sum_{n(\text{mod } p)} \left(\frac{n(n^2-l)}{p}\right)$, called Jacobsthal sums, are mentioned by Dickson (1952b), Chapter 6, page 253.

Chapter 11

Chapters 6, 7, and 8 of Dickson (1952b) are devoted to the problem of representing numbers as sums of two, three, and four squares.

Fermat's Conjecture is treated by Dickson (1952b), Chapter 26.

Chapters 12, 13, and 14

Dickson (1952b), Chapter 3, gives the history of partitions prior to 1920. An account of Ramanujan's contributions is given by Hardy (1940), Chapter 6. Alder (1969) and Andrews (1970) present up-to-date histories of partition identities.

The proof of Jacobi's Triple Product Identity (Theorem 13–8) was independently discovered by Andrews (1965) and Menon (1965).

The proof of the Rogers-Ramanujan identities is derived from Andrews (1966).

The exposition of Schur's theorem is an amalgamation of Andrews (1967) and (1968).

Chapter 15

Gauss's Circle Problem is described by Dickson (1952b), Chapter 6.

Dickson (1952a), Chapter 10, treats Dirichlet's Divisor Problem.

Appendix A

The proof that $\lim_{n\to\infty} p(n)^{1/n} = 1$ is taken from Andrews (1971).

Appendices B and C

Further developments of the theory of infinite series and products may be found in Bromwich (1959).

SUGGESTED READING

For further elementary topics in number theory the reader could do no better than to study the text by Hardy and Wright (1960). Advanced topics related to Parts I and III may be found in the books by Ayoub (1963), Grosswald (1966), and LeVeque (1956). *Vorlesungen über Zahlentheorie,* a classic treatise by Landau (1947), covers the theorems of our Chapter 15 in greater depth. For material on algebraic number theory, we refer the reader to the elementary text of Pollard (1950) and to the very advanced book of Weiss (1963). Finally, for further material on partitions, we suggest the work by Knopp (1970) as well as the survey articles by Alder (1969) and Andrews (1970).

BIBLIOGRAPHY

Alder, H. L. (1969). Partition identities—from Euler to the present. *Amer. Math. Monthly, 76,* 733–746.

Andrews, G. E. (1965). A simple proof of Jacobi's Triple Product Identity. *Proc. Amer. Math. Soc., 16,* 333–334.

Andrews, G. E. (1966). An analytic proof of the Rogers-Ramanujan-Gordon identities. *Amer. J. Math., 88,* 844–846.

Andrews, G. E. (1967). On Schur's second partition theorem. *Glasgow Math. J., 8,* 127–132.

Andrews, G. E. (1968). On partition functions related to Schur's second partition theorem. *Proc. Amer. Math. Soc., 19,* 441–444.

Andrews, G. E. (1969). On radix representation and the Euclidean algorithm. *Amer. Math. Monthly, 76,* 66–68.

Andrews, G. E. (1970). A polynomial identity which implies the Rogers-Ramanujan identities. *Scripta Mathematica, 28,* 297–305.

Andrews, G. E. (1971). A combinatorial proof of a partition function limit. *Amer. Math. Monthly, 78,* 276–278.

Ayoub, R. (1963). *An Introduction to the Analytic Theory of Numbers, Vol. 10.* American Mathematical Society Surveys, Providence.

Bromwich, T. J. I'A. (1959). *An Introduction to the Theory of Infinite Series.* Macmillan, London.

Carmichael, R. D. (1959). *The Theory of Numbers and Diophantine Analysis.* Dover, New York.

Dickson, L. E. (1952a). *History of the Theory of Numbers, Vol. 1.* Chelsea, New York.

Dickson, L. E. (1952b). *History of the Theory of Numbers, Vol. 2.* Chelsea, New York.

Golomb, S. W. (1956). Combinatorial proof of Fermat's "little" theorem. *Amer. Math. Monthly, 63,* 718.

Grosswald, E. (1966). *Topics from the Theory of Numbers.* Macmillan, New York.

Hardy, G. H. (1940). *Ramanujan.* Cambridge University Press, Cambridge.

Hardy, G. H., and Wright, E. M. (1960). *An Introduction to the Theory of Numbers,* 4th ed. Oxford University Press, Oxford.

Johnson, P. B. (1956). Congruences and card shuffling. *Amer. Math. Monthly, 63,* 718–719.

Knopp, M. I. (1970). *Modular Functions in Analytic Number Theory.* Markham, Chicago.

Knuth, D. E. (1969). *The Art of Computer Programming, Vol. 2.* Addison-Wesley, Reading.

Landau, E. (1947). *Vorlesungen über Zahlentheorie.* Chelsea, New York.

Lehmer, D. H. (1957). A low energy proof of the reciprocity law. *Amer. Math. Monthly, 64*, 103–106.

LeVeque, W. J. (1956). *Topics in Number Theory, Vol. 2.* Addison-Wesley, Reading.

Mamangakis, S. E. (1962). Remark on $\pi(x) = o(x)$. *Proc. Amer. Math. Soc., 13*, 664–665.

Menon, P. K. (1965). On Ramanujan's continued fraction and related identities. *J. London Math. Soc., 40*, 49–54.

Pollard, H. (1950). *The Theory of Algebraic Numbers* (Carus Monograph No. 9). Wiley, New York.

Riordan, J. (1958). *Introduction to Combinatorial Analysis.* Wiley, New York.

Uspensky, J. V., and Heaslet, M. A. (1939). *Elementary Number Theory.* McGraw-Hill, New York.

Weiss, E. (1963). *Algebraic Number Theory.* McGraw-Hill, New York.

HINTS AND ANSWERS TO SELECTED EXERCISES

SECTION 1-1

1. Since $1^2 = 1 = 1 \cdot (1+1) \cdot (2 \cdot 1+1)/6$, the formula is true for $n = 1$. Assuming that the formula is true for the integers $1,2,\ldots,k$, we see that

$$\begin{aligned}1^2+2^2+\ldots+(k+1)^2 &= (1^2+2^2+\ldots+k^2)+(k+1)^2\\ &= k(k+1)(2k+1)/6+(k+1)^2\\ &= (k+1)((k+1)+1)(2(k+1)+1)/6.\end{aligned}$$

Thus, whenever the formula is true for $1,2,\ldots,k$, it is also true for $k+1$.

2. Hint: Prove

$$1^3+2^3+\ldots+n^3 = \frac{1}{4}n^2(n+1)^2.$$

7. Since $1 = F_1 = 2-1 = F_3 - 1$, the formula is true for $n = 1$. Assuming that the formula is true for the integers $1,2,\ldots,k$ we see that

$$\begin{aligned}F_1+F_2+\ldots+F_{k+1} &= (F_1+F_2+\ldots+F_k)+F_{k+1}\\ &= (F_{k+2}-1)+F_{k+1} = (F_{k+2}+F_{k+1})-1\\ &= F_{k+3}-1.\end{aligned}$$

10. Since $F_2^2 - F_1F_3 = 1 - 1\cdot 2 = -1 = (-1)^1$, the formula is true for $n = 1$. Assuming that the formula is true for the integers $1,2,\ldots,k$, we see that

$$\begin{aligned}F_{k+2}^2 - F_{k+1}F_{k+3} &= F_{k+2}(F_{k+1}+F_k) - F_{k+1}(F_{k+2}+F_{k+1})\\ &= F_{k+2}F_k - F_{k+1}^2 = -(F_{k+1}^2 - F_kF_{k+2})\\ &= -(-1)^k = (-1)^{k+1}.\end{aligned}$$

17. Since $1 \cdot (1-1)(3 \cdot 1+2) = 0 = 24 \cdot 0$, the statement is true for $n=1$. Assuming that the formula is true for the integers $1,2,\ldots,k$, we see that

$$(k+1)((k+1)^2-1)(3(k+1)+2) = (k+1)(k^2+2k)(3k+5)$$
$$= k(k^2-1)(3k+2)+12k(k+1)^2.$$

Now by our assumption, 24 divides $k(k^2-1)(3k+2)$. Does 24 divide $12k(k+1)^2$? If k is even, then $k=2r$ and $12k(k+1)^2 = 24r(2r+1)^2$; if k is odd, then $k=2s+1$ and $12k(k+1)^2 = 48(2s+1)(s+1)^2$. Hence 24 does divide $12k(k+1)^2$, and the statement is proved by mathematical induction.

SECTION 1–2

1. 100, 112, 211.

3. 6.

6. Since we are representing a positive integer n, we know that $n > 0$. Since in any representation $a_i \leq k-1$, we see that

$$n = a_s k^s + a_{s-1}k^{s-1} + \ldots + a_0 \leq (k-1)(k^s + k^{s-1} + \ldots + 1)$$
$$= (k-1)(k^{s+1}-1)/(k-1) = k^{s+1}-1.$$

SECTION 2–1

1. Take $q = -|j|-1$.

6. Since a and b are odd integers, $a = 2r+1$, and $b = 2s+1$. Thus

$$a^2 - b^2 = (4r^2+4r+1) - (4s^2+4s+1) = 4(r-s)(r-s+1).$$

Now if $r-s$ is even, then $r-s=2m$ and $a^2-b^2 = 8m(2m+1)$; if $r-s$ is odd, then $r-s = 2n+1$ and $a^2-b^2 = 8(2n+1)(n+1)$. Thus in any case a^2-b^2 is divisible by 8 if a and b are odd integers.

SECTION 2–2

1. (a) 17; (c) 333; (d) 27.

5. (a) 150; (c) 81; (e) $n(n+1)$; (f) $(2n-1)(2n+1)$.

8. If $d = \text{g.c.d.}(a,b)$, then $d \mid a$ and $d \mid b$. Thus $d \mid (a+b)$ and $d \mid (a-b)$. Hence $d \mid \text{g.c.d.}(a+b, a-b)$, so $d \leq \text{g.c.d.}(a+b, a-b)$.

SECTION 2–3

1. (a) $y = 4 + 2t$, $x = -4 - 3t$; (c) no solutions; (e) $y = 21 + 5t$, $x = 21 + 4t$.

2. 5 apples and 4 pears.

4. Hint: As a special case, compute the area of the triangle with vertices (0,0),(b,a), and $(b,0)$, and then subtract both the area of the trapezoid with vertices $(x,0),(b,0),(b,a),(x,y)$ and the area of the triangle with vertices $(0,0),(x,y),(x,0)$.

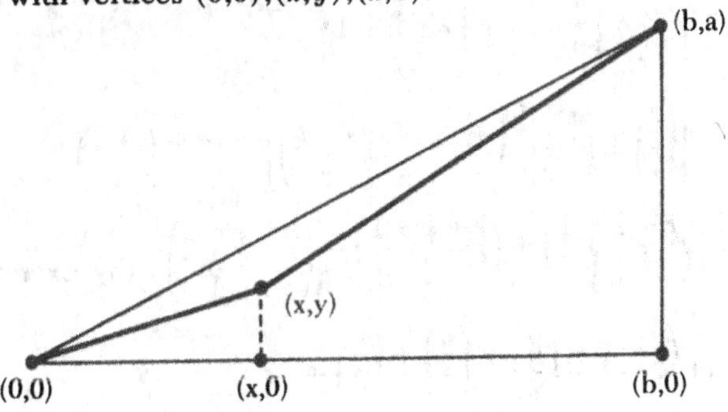

Then consider the other possible ways the triangle (0,0), (b,a), (x,y) could be positioned in the plane.

7. $1/(a^2 + b^2)^{1/2}$.

8. $(a^2 + b^2)^{1/2}/\text{g.c.d.}(a,b)$.

SECTION 2–4

1. $60 = 2 \cdot 30 = 6 \cdot 10$.

3. Hint: In the factorization of a, collect all repetitions of each prime factor as that prime raised to a power. For example, $36 = 2 \cdot 2 \cdot 3 \cdot 3 = 2^2 \cdot 3^2$.

6. (a) 11; (c) 17; (f) 1.

9. (a) 750; (e) 1221; (f) p^2qr.

11. g.c.d.(39,102,75) = 3; l.c.m.(39,102,75) = 33150.

12. Hint: use Exercises 5, 7, and 10.

SECTION 3–1

2. The total number of subsets of a set of n elements is 2^n by Exercise 1. On the other hand, there is 1 subset with no elements; there are $\binom{n}{1}$ subsets with one element, $\binom{n}{2}$ with two elements, and so on. This alternative way of counting the subsets of S establishes the formula.

5. Since $\binom{r}{r}+\binom{r+1}{r}=1+r+1=r+2=\binom{r+2}{r+1}$, the formula is true for $m=1$. Assuming that the formula is true when m is any of the integers $1,2,\ldots,k$, we see that

$$\binom{r}{r}+\binom{r+1}{r}+\ldots+\binom{r+k+1}{r}$$

$$=\left\{\binom{r}{r}+\binom{r+1}{r}+\ldots+\binom{r+k}{r}\right\}+\binom{r+k+1}{r}$$

$$=\binom{r+k+1}{r+1}+\binom{r+k+1}{r}=\binom{r+k+2}{r+1} \quad \text{(by Exercise 3).}$$

7. Let $h_n=1+\binom{n}{2}+\binom{n}{4}+\binom{n}{6}+\ldots,$

and

$$k_n=\binom{n}{1}+\binom{n}{3}+\binom{n}{5}+\binom{n}{7}+\ldots.$$

Then

$$h_n+k_n=1+\binom{n}{1}+\binom{n}{2}+\ldots+\binom{n}{n}=2^n \quad \text{(by Exercise 2),}$$

and

$$h_n-k_n=1-\binom{n}{1}+\binom{n}{2}-\ldots+(-1)^n\binom{n}{n}=(-1)^{n-1}\binom{n-1}{n}=0$$

(by Exercise 6). Now we have two equations for h_n and k_n. Solving these equations, we find that $h_n=k_n=2^{n-1}$.

9. Since $\binom{p}{a}=\dfrac{p(p-1)\ldots(p-a+1)}{a!}$ is an integer, we see that $a!\mid p(p-1)\ldots(p-a+1)$. But since $a<p$, we must have g.c.d.$(a!,p)=1$; therefore, by Theorem 2–3, $a!\mid(p-1)\ldots(p-a+1)$. Hence $\binom{p}{a}=p\cdot\dfrac{(p-1)\ldots(p-a+1)}{a!}$.

12. $$\sum_{s=0}^{n} \binom{n}{s} (-1)^s a_s = \sum_{s=0}^{n} \binom{n}{s} (-1)^s \sum_{r=0}^{s} \binom{s}{r} b_r$$

$$= \sum_{\substack{s=0\\ s\geqq r}}^{n} \sum_{r=0}^{n} \binom{n}{s}\binom{s}{r} (-1)^s b_r$$

$$= \sum_{r=0}^{n} b_r \sum_{s=r}^{n} \frac{n!}{s!(n-s)!} \frac{s!}{r!(s-r)!} (-1)^s$$

$$= \sum_{r=0}^{n} \frac{n!}{r!(n-r)!} b_r \sum_{s=r}^{n} \frac{(n-r)!(-1)^s}{(n-s)!(s-r)!}$$

$$= \sum_{r=0}^{n} \binom{n}{r} b_r \sum_{s=r}^{n} \binom{n-r}{s-r} (-1)^s$$

$$= \sum_{r=0}^{n} \binom{n}{r} b_r \sum_{s=0}^{n-r} \binom{n-r}{s} (-1)^{s+r}.$$

Now if $n - r > 1$, then by Exercise 10 (with $x = 1$, $y = -1$),

$$0 = (1-1)^{n-r} = \sum_{s=0}^{n-r} \binom{n-r}{s} (-1)^s.$$

Consequently, the last expression in our string of equations has only one nonzero term, namely that in which $n = r$, and that term is $(-1)^n b_n$.

13. Clearly the assertion is true for $n = 1$ and 2. Suppose that it is true for the integers $1,2,\ldots,k$; then

$$1 + \binom{k+1-2}{1} + \binom{k+1-3}{2} + \binom{k+1-4}{3} + \ldots$$

$$= 1 + \left\{\binom{k-2}{1} + \binom{k-2}{0}\right\} + \left\{\binom{k-3}{2} + \binom{k-3}{1}\right\}$$

$$+ \left\{\binom{k-4}{3} + \binom{k-4}{2}\right\} + \ldots$$

$$= \left\{1 + \binom{k-2}{1} + \binom{k-3}{2} + \binom{k-4}{3} + \ldots\right\}$$

$$+ \left\{1 + \binom{k-3}{1} + \binom{k-4}{2} + \ldots\right\} = F_k + F_{k-1} = F_{k+1}.$$

SECTION 3–2

1. By Fermat's little theorem, $p \mid n(n^{p-1}-1)$. Therefore, by Corollary 2–3, $p \mid n^{p-1}-1$.

3. Hint: prove that $n^5 \equiv n \pmod 5$ and $n^5 \equiv n \pmod 2$.

5. Hint: use Exercise 1.

SECTION 3–3

1. Any prime smaller than p divides $(p-1)!$ and so does not divide $(p-1)!+1$. On the other hand, by Wilson's theorem, $p \mid ((p-1)!+1)$.

2. The expression $(n-1)!+1$ is equal to 2 for $n=1$ and 2, to 3 for $n=3$, to 7 for $n=4$, and to 25 for $n=5$. For $n>5$, $(n-1)!$ is divisible by 10 and so $(n-1)!+1$ is not.

SECTION 3–4

1. $$\frac{x}{(1-x)^2} = x\frac{d}{dx}(1-x)^{-1} = x\frac{d}{dx}(1+x+x^2+x^3+\ldots)$$
$$= x(1+2x+3x^2+4x^3+\ldots)$$
$$= x+2x^2+3x^3+4x^4+\ldots.$$

4. $$\frac{1}{(1-x)^3} = \frac{1}{2}\frac{d^2}{dx^2}(1+x+x^2+x^3+\ldots)$$
$$= \frac{1}{2}(2+3\cdot 2x+4\cdot 3x^2+5\cdot 4x^3+\ldots)$$
$$= 1+\binom{3}{2}x+\binom{4}{2}x^2+\binom{5}{2}x^3+\ldots.$$

6. $$(1-2x)^{-1} = 1+2x+(2x)^2+(2x)^3+\ldots$$
$$= 1+2x+4x^2+8x^3+\ldots.$$

7. Hint: decompose the expression $\dfrac{x}{(1-bx)(1-bx-x)}$ by partial fractions.

SECTION 4–1

1. (a) 2; (b) 3; (c) 4. (There are other correct answers.)

3. $(a_0x^r + a_1x^{r-1} + \ldots + a_r) - (a_0y^r + a_1y^{r-1} + \ldots + a_r) = a_0(x^r - y^r) + a_1(x^{r-1} - y^{r-1}) + \ldots + a_{r-1}(x - y)$. Hence we need only show that $m \mid (x^r - y^r)$ for each positive integer r. This is clear since $m \mid (x-y)$, and $x^r - y^r = (x-y)(x^{r-1} + x^{r-2}y + \ldots + xy^{r-2} + y^{r-1})$.

4. This is just a special case of the cancellation law.

7. (b) yes; (c) no; (e) yes.

SECTION 4–2

1. All are complete residue systems modulo 11.

2. (a) and (d) are reduced residue systems modulo 18.

4. Yes, every number counted by $w(n)$ is also counted by $\phi(n)$, and since g.c.d.$(1,n) = 1$, we see that there is always a non-prime integer (namely 1) smaller than n that is relatively prime to n.

$$w(n) = \phi(n) - 1 \text{ for } n = 1,2,3,4,6,8,12,18,24,30.$$

SECTION 4–3

1. 3, 6, 12.

SECTION 5–1

1. (a) 7; (b) 5, 20; (c) 3, 8, 13.

3. (a) 6; (b) 4; (c) 10.

SECTION 5–2

1.

1	2	3	4	5	6	7	8	9	10	11	12
27	15	3	30	18	6	33	21	9	36	24	12

3. $(a-1)(1+a+\ldots+a^{\phi(m)-1})=a^{\phi(m)}-1$; now $m \mid a^{\phi(m)}-1$ and g.c.d.$(a-1,m)=1$. Hence $m \mid (1+a+\ldots+a^{\phi(m)-1})$.

5. $41^{75} \equiv (-1)^{75} \equiv -1 \equiv 2 \pmod 3$.

7. Hint: $10^n \equiv 1^n \equiv 1 \pmod 9$.

8. If $x=a^{p-2}b$, then $ax=a^{p-1}b \equiv 1 \cdot b \equiv b \pmod p$.

9. Since $p \equiv 1 \pmod 4$, we may write $p=4t+1$. Hence $(4t)! \equiv -1 \pmod p$, and

$$\begin{aligned}(4t)! &= (2t)!(2t+1)(2t+2)\ldots(2t+2t)\\ &= (2t)!(p-2t)(p-(2t-1))\ldots(p-1)\\ &\equiv (2t)!(-2t)(-(2t-1))\ldots(-1) \pmod p\\ &\equiv (2t)!(-1)^{2t}(2t)! \equiv (2t)!^2 \pmod p,\end{aligned}$$

and $\quad 2t=\frac{1}{2}(p-1)$.

10. (a) 720; (b) 40320.

14. There are no primes in this interval since $m!+j$ is divisible by j for $2 \leq j \leq m$.

17. $16!+1=20922789888001 = 17 \cdot 1230752346353$. This is certainly a candidate for the most inefficient possible test for determining whether 1093 is a prime.

20. (a) $\Phi_n(\Phi_n-2)=(2^{2^n}+1)(2^{2^n}-1)=(2^{2^n})^2-1=2^{2^{n+1}}-1$.

(b) If $b=ca$, then

$$2^b-1=(2^a)^c-1=(1+2^a+\ldots+2^{a(c-1)})(2^a-1).$$

(c) Since $n+1 \leq 2^n$ (by Corollary 1–1), $2^{n+1} \mid 2^{2^n}$; hence, by (b),

$$(2^{2^{n+1}}-1) \mid (2^{2^{2^n}}-1).$$

(d) $(2^{2^{2^n}+1}-2)=2(2^{2^{2^n}}-1)$.

(e) $\Phi_n \mid (2^{2^{n+1}}-1) \mid (2^{2^{2^n}}-1) \mid (2^{2^{2^n}+1}-2)=2^{\Phi_n}-2$.

22. By Exercise 21, $2^l \equiv \pm 1 \pmod p$, and $l! \equiv \pm 1 \pmod p$ by Exercise 15. Hence

$$\begin{aligned}-1 \equiv (2l)! &\equiv 2^l l! \cdot 1 \cdot 3 \cdot 5 \ldots (2l-1) \pmod p\\ &\equiv \pm 1 \cdot 3 \cdot 5 \ldots (2l-1) \pmod p.\end{aligned}$$

SECTION 5–3

1. (a) All integers congruent to 23 modulo 30.

4. Hint: in Theorem 5–4, choose m_i as the square of the ith prime. Then consider the congruences

$$x \equiv 0 \pmod{2^2}$$
$$x+1 \equiv 0 \pmod{3^2}$$
$$\cdots$$
$$x+n-1 \equiv 0 \pmod{p_n^2}.$$

6. 29.

SECTION 5–4

1. (a) No solutions; (b) all integers congruent to 7 modulo 13; (c) all integers congruent to 4 modulo 7.

7. Both sides of the congruence are congruent to zero for $x = 1,2,\ldots,p-1$. Since $(x^{p-1}-1)-(x-1)(x-2)\ldots(x-p+1)$ is a polynomial of degree less than $p-1$, it must be congruent to zero modulo p for each integer x.

8. Wilson's theorem.

SECTION 6–1

1. If $m = p^j q^k \ldots r^l$ is the prime factorization of m and $j \geq 2$, then by (6–1–2), $p \mid p(p^{j-1}-p^{j-2}) = p^j - p^{j-1} = \phi(p^j) \mid \phi(m)$. Hence $p \mid \phi(m)$ and $p \nmid m-1$. Therefore $\phi(m) \nmid m-1$.

2. If m has only two distinct prime factors and $\phi(m) \mid m-1$, then by Exercise 1, $m = pq$ where p and q are distinct primes. Thus

$$\frac{m-1}{\phi(m)} = \frac{pq-1}{(p-1)(q-1)} = \frac{pq-p-q+1+p+q-2}{(p-1)(q-1)}$$

$$= \frac{(p-1)(q-1)}{(p-1)(q-1)} + \frac{p-1}{(p-1)(q-1)} + \frac{q-1}{(p-1)(q-1)}$$

$$= 1 + \frac{1}{q-1} + \frac{1}{p-1}.$$

Since we assume that $\phi(m) \mid m-1$, then $(m-1)/\phi(m)$ must be an integer; however, $1 < 1 + \frac{1}{p-1} + \frac{1}{q-1} \leq 3$ if p and q are primes. Thus we must have

$$\frac{1}{p-1} + \frac{1}{q-1} = 1 \text{ or } 2,$$

and this is possible only for $p = q = 2$ or $p = q = 3$. Finally, we observe that each of these solutions is inadmissible because p and q are distinct primes.

5. $\phi(n) = n \prod_{p \mid n} \left(1 - \frac{1}{p}\right) \geq n \prod_{p \mid n} \left(1 - \frac{1}{2}\right) = n \prod_{p \mid n} \frac{1}{2} = n2^{-r}.$

7. Yes. If Goldbach's conjecture is true, then for each n there exist primes p and q such that

$$2n + 2 = p + q = \phi(p) + 1 + \phi(q) + 1.$$

Hence

$$2n = \phi(p) + \phi(q).$$

8. Hint: if $nx_0 + my_0 \equiv nx_1 + my_1 \pmod{mn}$, where g.c.d.$(x_0, m) =$ g.c.d.$(x_1, m) =$ g.c.d.$(y_0, n) =$ g.c.d.$(y_1, n) = 1$, then

$$n(x_0 - x_1) \equiv m(y_1 - y_0) \pmod{mn}.$$

By Theorem 5–1, this is possible only if $n \mid m(y_1 - y_0)$, but g.c.d.$(n, m) = 1$; hence $n \mid (y_1 - y_0)$. A similar argument shows that $m \mid (x_0 - x_1)$.

12. (a) $\phi(4s + 2) = \phi(2(2s + 1)) = \phi(2)\,\phi(2s + 1) = \phi(2s + 1)$.

(b) $\phi(1) = \phi(2) = 1$, $\phi(3) = \phi(4) = \phi(6) = 2$, $\phi(5) = \phi(10) = \phi(8) = \phi(12) = 4$, $\phi(7) = \phi(9) = \phi(14) = \phi(18) = 6$, $\phi(11) = \phi(22) = 10$, $\phi(13) = \phi(21) = 12$, $\phi(15) = \phi(16) = \phi(20) = 8$, $\phi(17) = \phi(32) = 16$, $\phi(19) = \phi(38) = 18$.

13. $\phi(11^n) = 11^n - 11^{n-1} = 11^{n-1} \cdot 10.$

14. $\phi(2^{2n+1}) = 2^{2n+1} - 2^{2n} = 2^{2n}(2 - 1) = 2^{2n} = (2^n)^2.$

SECTION 6–2

1. We say that x and y are *complementary divisors* of n if $xy=n$. Thus we may separate the divisors of n into pairs of complementary divisors. This means that there will be an even number of divisors of n except when one (or any odd number) of the pairs of complementary divisors does not contain two distinct integers. This occurs only when $x=y$ or $x^2=n$. Hence if n is a perfect square, then n has an odd number of divisors. Otherwise n has an even number of divisors.

4. Hint: if $x+(x+1)+\ldots+(x+r)=n$, then

$$(r+1)x+\frac{1}{2}r(r+1)=n.$$

7. Hint: since $1+p+\ldots+p^n=\dfrac{p^{n+1}-1}{p-1}$, $1+2p+3p^2+\ldots$ $+np^{n-1}=\dfrac{d}{dp}(1+p+\ldots+p^n)=\dfrac{d}{dp}\dfrac{p^{n+1}-1}{p-1}$.

9. $2n=\sigma(n)=\sum_{d\mid n} d=\sum_{d\mid n}\dfrac{n}{d}=n\sum_{d\mid n}\dfrac{1}{d}$; hence $2=\sum_{d\mid n}\dfrac{1}{d}$.

12. If $n=p$, a prime, then

$$\frac{\phi(n)\sigma(n)+1}{n}=\frac{(p-1)(p+1)+1}{p}=\frac{p^2-1+1}{p}=p.$$

If $n=p^jm$, where $j\geq 2$ and g.c.d.$(p,m)=1$, then

$$\frac{\phi(n)\sigma(n)+1}{n}=\frac{\phi(p^j)\phi(m)\sigma(n)+1}{p^jm}=\frac{p(p^{j-1}-p^{j-2})\phi(m)\sigma(n)+1}{p^jm}.$$

Hence p is not a factor of the numerator, but it is a factor of the denominator.

SECTION 6–4

1. Since $f(n)=n^k$ is obviously multiplicative, $\sum_{d\mid n} f(d)=\sum_{d\mid n} d^k=$ $\sigma_k(n)$ is also multiplicative by Theorem 6–7.

2. Since both $f(n)$ and $\mu(n)$ are multiplicative, so is $f(n)\,\mu(n)$, and thus by Theorem 6–7, so is

$$F(n) = \sum_{d\mid n} \mu(d) f(d).$$

Thus if $n = p_1^{\alpha_1} p_2^{\alpha_2} \ldots p_r^{\alpha_r}$, then

$$\begin{aligned} F(n) &= F(p_1^{\alpha_1}) F(p_2^{\alpha_2}) \ldots F(p_r^{\alpha_r}) \\ &= (1 - f(p_1))(1 - f(p_2)) \ldots (1 - f(p_r)) \\ &= \prod_{p\mid n} (1 - f(p)). \end{aligned}$$

7. Since $\mu(n)$ and $\phi(n)$ are multiplicative and $\phi(n) \neq 0$, $\mu^2(n)/\phi(n)$ is multiplicative and thus $\sum_{d\mid n} \mu^2(d)/\phi(d) = G(n)$ is also multiplicative by Theorem 6–7. Hence, if $n = p_1^{\alpha_1} p_2^{\alpha_2} \ldots p_r^{\alpha_r}$, then

$$\begin{aligned} G(n) &= G(p_1^{\alpha_1}) G(p_2^{\alpha_2}) \ldots G(p_r^{\alpha_r}) \\ &= \left(1 + \frac{1}{\phi(p_1)}\right) \ldots \left(1 + \frac{1}{\phi(p_r)}\right) = \frac{p_1}{p_1 - 1} \cdots \frac{p_r}{p_r - 1} \\ &= \frac{1}{1 - \dfrac{1}{p_1}} \cdots \frac{1}{1 - \dfrac{1}{p_r}} = \frac{n}{n\left(1 - \dfrac{1}{p_1}\right) \ldots \left(1 - \dfrac{1}{p_r}\right)} = \frac{n}{\phi(n)}. \end{aligned}$$

13. $$\begin{aligned} \sum_{n=1}^{[x]} \mu(n) f(x/n) &= \sum_{n=1}^{[x]} \mu(n) \sum_{m=1}^{[x/n]} g(x/mn) \\ &= \sum_{\substack{mn \leqq x \\ m \geqq 1, n \geqq 1}} \mu(n) g(x/mn) = \sum_{k=1}^{[x]} g(x/k) \sum_{mn=k} \mu(n) \\ &= g(x). \end{aligned}$$

14. Let a_i be the ith divisor of n. Then

$$\frac{a_1 + a_2 + \ldots + a_{d(n)}}{d(n)} \geq (a_1 a_2 \ldots a_{d(n)})^{1/d(n)} = \prod_{d\mid n} d^{1/d(n)}.$$

SECTION 7–1

1. (a) If $r = \operatorname{ind}_g a$ and $s = \operatorname{ind}_g b$, then

$$g^r \equiv a \equiv b \equiv g^s \pmod{m}.$$

Hence, since g belongs to the exponent $\phi(m)$ modulo m, we see that $\phi(m) \mid r-s$; that is, $\text{ind}_g a \equiv \text{ind}_g b \pmod{\phi(m)}$.

2. We only need a table for the positive integers less than 17:

a	1	2	3	4	5	6	7	8	9	10	11	12	13	14	15	16
$\text{ind}_3 a$	16	14	1	12	5	15	11	10	2	3	7	13	4	9	6	8

3. $\text{ind}_3 9 + \text{ind}_3 x \equiv \text{ind}_3 11 \pmod{16}$

$$2 + \text{ind}_3 x \equiv 7 \pmod{16}$$

$$\text{ind}_3 x \equiv 5 \pmod{16}.$$

Hence x must be congruent to 5 modulo 17.

6. 2,3 (mod 5); 2,5 (mod 9); 2,6,7,8 (mod 11); 2,6,7,11 (mod 13); none (mod 15).

SECTION 7-2

1. Since $j = h \cdot (k/\text{g.c.d.}(h,k)) = k \cdot (h/\text{g.c.d.}(h,k))$, we see that both h and k divide j. Hence

$$a^j \equiv 1 \pmod{m}, \text{ and } a^j \equiv 1 \pmod{n};$$

therefore, $m \mid a^j - 1$ and $n \mid a^j - 1$. However, $\text{g.c.d.}(m,n) = 1$, and so $mn \mid a^j - 1$, or $a^j \equiv 1 \pmod{mn}$.

3. If $n \geq 3$, and $\text{g.c.d.}(a, 2^n) = 1$, then by Exercise 2,

$$a^{2^{n-2}} \equiv 1 \pmod{2^n},$$

but $\phi(2^n) = 2^{n-1} > 2^{n-2}$. Thus, no integer belongs to $\phi(2^n)$ modulo 2^n, and so there are no primitive roots modulo 2^n.

7. $(g+p)^{p-1} = g^{p-1} + \binom{p-1}{1} g^{p-2} p + \text{terms involving } p^2$

$$\equiv g^{p-1} + (p-1) g^{p-2} p \pmod{p^2}$$

$$\equiv 1 - g^{p-2} p \pmod{p^2}.$$

If this last expression were congruent to 1 modulo p^2, it would imply that $p \mid g$, an impossibility.

9. Hint: $\phi(p^m) = p^m - p^{m-1}$.

10. By assumption, the assertion is true for $m = 2$. Assume that it is true for the integers $2,3,\ldots,k$. Now

$$g^{\phi(p^{k-1})} = g^{p^{k-2}(p-1)} \equiv 1 \pmod{p^{k-1}}.$$

Hence since the assertion is assumed true for k, we must have $g^{p^{k-2}(p-1)} = 1 + rp^{k-1}$, where $p \nmid r$. Therefore,

$$\begin{aligned} g^{p^{k-1}(p-1)} = (g^{p^{k-2}(p-1)})^p &= (1 + rp^{k-1})^p \\ &= 1 + \binom{p}{1} rp^{k-1} + \text{terms involving } p^{k+1} \\ &\equiv 1 + rp^k \pmod{p^{k+1}}. \end{aligned}$$

Hence,

$$g^{p^{k-1}(p-1)} \not\equiv 1 \pmod{p^{k+1}},$$

since $p \nmid r$.

15. One modulo 6, two modulo 7, none modulo 8, two modulo 9, two modulo 10.

SECTION 8–1

1. See Exercise 20 of Section 5–2.

2. If $n > m$, $d \mid \Phi_m$, and $d \mid \Phi_n$, then by Exercise 1, $d \mid 2$. Hence d is either 2 or 1, and since Φ_n is odd, $d = 1$.

3. Let p_n be a prime factor of Φ_n. Then $p_1, p_2, p_3, \ldots$ must be an infinite sequence of distinct primes since no Φ_n has any prime factor in common with any other Φ_m.

4. Hint: Assume that there are exactly r primes $q_1, q_2, \ldots, q_r$ that are congruent to 5 modulo 6. Examine the prime factors of $6q_1q_2\ldots q_r - 1$.

8. $a_2 = 3$, $a_3 = 5$, $a_4 = 7$, $a_5 = 11$, $a_6 = 13$, a_n is the nth prime.

9. $2 = p_1 < 2^{2^1} + 1 = 5$. Assume the assertion for $1, 2, \ldots, k$. Then

$$p_{k+1} \le p_1 p_2 \ldots p_k + 1 \le 2^{2^1} \cdot 2^{2^2} \ldots 2^{2^k} + 1 = 2^{2+2^2+\ldots+2^k} + 1$$
$$= 2^{2^{k+1}-2} + 1 < 2^{2^{k+1}} + 1.$$

12. $\theta(x) = \sum_{p \le x} \log p \le \sum_{p \le x} \log x = \log x \sum_{p \le x} 1 = \pi(x) \log x.$

13. $\phi(M) = (2-1)(3-1)(5-1)\ldots(p-1) > 1$. Thus there exist integers not exceeding M having no prime factors among $2,3,5,\ldots,p$.

14. $x = 41$.

15. $20! = 2^{18}3^8 5^4 7^2 11 \cdot 13 \cdot 17 \cdot 19$.

SECTION 8–2

1. $$\pi(125x) - \pi(x) > \frac{\log 2}{4} \frac{125x}{\log 125x} - 30 \log 2 \frac{x}{\log x}$$
$$= x \log 2 \left(\frac{31.25}{\log 125x} - \frac{30}{\log x} \right),$$

and if x is sufficiently large, $\dfrac{\log x}{\log 125x} = \dfrac{\log x}{\log x + \log 125} > \dfrac{30}{31.25}$. Consequently, for x sufficiently large, $\pi(125x) - \pi(x) > 0$.

2. Since $\frac{2}{3} n < p \le n$, we see that $\frac{4}{3} n < 2p \le 2n$, and $2n < 3p$. Hence, there is exactly one term among $(n+1),\ldots,2n$ that is divisible by p (namely $2p$), and there is exactly one term among $1,2,\ldots,n$ that is divisible by p (namely p itself). Hence $\binom{2n}{n} = 2n(2n-1)\ldots(n+1)/n!$ has p appearing *once* in the denominator and *once* in the numerator. Therefore, since these factors cancel, we see that p is not a factor of $\binom{2n}{n}$.

SECTION 9–1

1. (a) $2^2 = 4 \equiv -1 \pmod 5$; therefore, 2 is not a quadratic residue modulo 5.

(b) $4^3 = 64 \equiv 1 \pmod 7$; therefore, 4 is a quadratic residue modulo 7.

(c) $3^5 = 243 \equiv 1 \pmod{11}$; therefore, 3 is a quadratic residue modulo 11.

(d) $6^6 = 46656 \equiv -1 \pmod{13}$; therefore, 6 is not a quadratic residue modulo 13.

SECTION 9–2

1. Suppose $c = p_1 p_2 \ldots p_r$; then

$$\left(\frac{ab}{c}\right) = \left(\frac{ab}{p_1}\right)\left(\frac{ab}{p_2}\right)\cdots\left(\frac{ab}{p_r}\right)$$
$$= \left(\frac{a}{p_1}\right)\left(\frac{b}{p_1}\right)\left(\frac{a}{p_2}\right)\left(\frac{b}{p_2}\right)\cdots\left(\frac{a}{p_r}\right)\left(\frac{b}{p_r}\right)$$
$$= \left(\frac{a}{p_1}\right)\left(\frac{a}{p_2}\right)\cdots\left(\frac{a}{p_r}\right)\left(\frac{b}{p_1}\right)\left(\frac{b}{p_2}\right)\cdots\left(\frac{b}{p_r}\right) = \left(\frac{a}{c}\right)\left(\frac{b}{c}\right).$$

SECTION 9–3

1. If $c = p_1 p_2 \ldots p_r$, then

$$\left(\frac{-1}{c}\right) = \left(\frac{-1}{p_1}\right)\left(\frac{-1}{p_2}\right)\cdots\left(\frac{-1}{p_r}\right)$$
$$= (-1)^{\frac{1}{2}(p_1-1)}(-1)^{\frac{1}{2}(p_2-1)}\cdots(-1)^{\frac{1}{2}(p_r-1)}$$
$$= \begin{cases} 1 & \text{if an even number of the } p_i \text{ are congruent to } 3 \pmod 4 \\ -1 & \text{if an odd number of the } p_i \text{ are congruent to } 3 \pmod 4 \end{cases}$$
$$= \begin{cases} 1 & \text{if } c \equiv 1 \pmod 4 \\ -1 & \text{if } c \equiv -1 \pmod 4 \end{cases}$$
$$= (-1)^{\frac{1}{2}(c-1)}.$$

4. The least residues modulo 19 of 17, 34, 51, 68, 85, 102, 119, 136, and 153 are −2, −4, −6, −8, 9, 7, 5, 3, and 1 respectively. Hence, $\left(\frac{17}{19}\right) = (-1)^{\mu} = (-1)^4 = 1$. Therefore, 17 is a quadratic residue modulo 19.

SECTION 9–4

1. $\left(\frac{17}{29}\right)=\left(\frac{29}{17}\right)=\left(\frac{12}{17}\right)=\left(\frac{4}{17}\right)\left(\frac{3}{17}\right)=\left(\frac{3}{17}\right)=\left(\frac{17}{3}\right)=\left(\frac{2}{3}\right)=-1.$
Hence, $x^2 \equiv 17 \pmod{29}$ has no solutions.

2. This problem is equivalent to $x^2 \equiv 4 \pmod{23}$, and $x=2$ is an obvious solution.

4. Hint: $(x+a)^2+b \equiv x^2+5x-12 \pmod{31}$ holds provided that $2a \equiv 5 \pmod{31}$, and $a^2+b \equiv -12 \pmod{31}$. Hence, we may take $a=18$ and $b=5$. To finish we need only solve the congruence $y^2 \equiv -5 \pmod{31}$.

5. If $p=12k+1$, then $p \equiv 1 \pmod 4$, and $\left(\frac{3}{p}\right)=\left(\frac{p}{3}\right)=\left(\frac{12k+1}{3}\right)=\left(\frac{1}{3}\right)=1$. The other cases are handled similarly.

SECTION 10–1

4. Hint: $\Upsilon(f)=\sum_{n=0}^{p-1}\left\{1+\left(\frac{f(n)}{p}\right)\right\}$.

5. The number of solutions is

$$p+\sum_{n=0}^{p-1}\left(\frac{n^2+3n+2}{p}\right)=p+\sum_{n=0}^{p-1}\left(\frac{(n+1)(n+2)}{p}\right)=p-1.$$

6. Hint:

$$\frac{1}{2}\left\{1+\left(\frac{d}{p}\right)\right\}|\mu(d)|=\begin{cases}1 & \text{if } d \text{ is a square-free quadratic residue modulo } p\\ 0 & \text{otherwise.}\end{cases}$$

SECTION 10–2

2. The number of solutions is

$$p+\sum_{n=0}^{p-1}\left(\frac{n^3+3n^2+2n}{p}\right)=p+\sum_{n=0}^{p-1}\left(\frac{n(n+1)(n+2)}{p}\right)$$

$$=p+S(1)=\begin{cases}p & \text{if } p\equiv 3 \pmod 4\\ >p-2\sqrt{p} & \text{if } p\equiv 1 \pmod 4).\end{cases}$$

SECTION 11–1

1. $274625 = 5^3 13^3 = 5^2 5 \cdot 13^2 13 = 65^2(2^2+1^2)(3^2+2^2)$
$= 65^2(8^2+1^2) = 520^2 + 65^2.$

2. $333 = 9 \cdot 37 = 3^2(6^2+1^2) = 18^2 + 3^2.$

SECTION 11–2

9. Suppose that $4^a(8m+7) = x^2+y^2+z^2$ where $a > 0$; then x, y, and z must all be even. Otherwise, $x^2+y^2+z^2 \not\equiv 0 \pmod 4$. Hence, $x = 2x_0$, $y = 2y_0$, and $z = 2z_0$; therefore,

$$4^a(8m+7) = 4(x_0^2+y_0^2+z_0^2),$$

or

$$4^{a-1}(8m+7) = x_0^2+y_0^2+z_0^2.$$

We may continue this process until we reduce the exponent of 4 to zero. Consequently, $4^a(8m+7)$ is a sum of three squares only if $8m+7$ is. If $8m+7 = x^2+y^2+z^2$, then

$$x^2+y^2+z^2 \equiv 7 \pmod 8.$$

However, since a perfect square is congruent to one of 0, 1, or 4 modulo 8, we see that

$$x^2+y^2+z^2 \equiv 0, 1, 2, 3, 4, 5, \text{ or } 6 \pmod 8.$$

Thus $8m+7$ is not equal to the sum of three squares.

SECTION 12–2

1. (a)

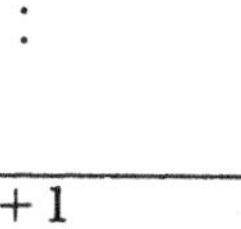

```
. . . . . .
. . . .
. . .
. .
.
.
```

2.

6	1 + 1 + 1 + 1 + 1 + 1
5 + 1	2 + 1 + 1 + 1 + 1
4 + 2	2 + 2 + 1 + 1
4 + 1 + 1	3 + 1 + 1 + 1
3 + 3	2 + 2 + 2
3 + 2 + 1	3 + 2 + 1
3 + 1 + 1 + 1	4 + 1 + 1
2 + 2 + 2	3 + 3
2 + 2 + 1 + 1	4 + 2

SECTION 12–3

1.

9	9
7+1+1	7+2
5+3+1	5+3+1
5+1+1+1+1	5+4
3+3+3	6+3
3+3+1+1+1	6+2+1
3+1+1+1+1+1+1	4+3+2
1+1+1+1+1+1+1+1+1	8+1

SECTION 12–4

10. Hint: the answer is *not* the set of positive integers congruent to 1, 2, or 3 modulo 6.

SECTION 13–1

1. Hint: Separate the partitions into two classes: (1) those that have m as a summand, and (2) those that do not.

4. Hint: See the preceding hint.

7. $\prod_{n=0}^{\infty} (1+q^{7n+1})(1+q^{7n+2})(1+q^{7n+4})$

$$= \prod_{n=0}^{\infty} \frac{(1-q^{14n+2})(1-q^{14n+4})(1-q^{14n+8})}{(1-q^{7n+1})(1-q^{7n+2})(1-q^{7n+4})}$$

$$= \prod_{n=0}^{\infty} \frac{(1-q^{14n+2})(1-q^{14n+4})(1-q^{14n+8})}{(1-q^{14n+1})(1-q^{14n+8})(1-q^{14n+2})(1-q^{14n+9})(1-q^{14n+4})(1-q^{14n+11})}$$

$$= \prod_{n=0}^{\infty} \frac{1}{(1-q^{14n+1})(1-q^{14n+9})(1-q^{14n+11})}.$$

SECTION 13–2

4. Hint: in the result of Exercise 3, first replace z by bz, and then replace a by a/b.

6.

$$\sum_{n=0}^{\infty} \frac{q^{\frac{1}{2}n(n+1)}(1-a)(1-aq)\ldots(1-aq^{n-1})}{(1-q)(1-q^2)\ldots(1-q^n)}$$

$$=\sum_{n=0}^{\infty}\frac{q^{\frac{1}{2}n(n+1)}}{(1-q)(1-q^2)\dots(1-q^n)}\frac{\prod_{m=0}^{\infty}(1-aq^m)}{\prod_{m=0}^{\infty}(1-aq^{m+n})}$$

$$=\prod_{m=0}^{\infty}(1-aq^m)\sum_{n=0}^{\infty}\frac{q^{\frac{1}{2}n(n+1)}}{(1-q)(1-q^2)\dots(1-q^n)}\sum_{r=0}^{\infty}\frac{a^r q^{rn}}{(1-q)(1-q^2)\dots(1-q^r)}$$

7. Continuing from the preceding answer:

$$=\prod_{m=0}^{\infty}(1-aq^m)\sum_{r=0}^{\infty}\frac{a^r}{(1-q)(1-q^2)\dots(1-q^r)}\sum_{n=0}^{\infty}\frac{q^{\frac{1}{2}n(n+1)+rn}}{(1-q)(1-q^2)\dots(1-q^n)}$$

$$=\prod_{m=0}^{\infty}(1-aq^m)\sum_{r=0}^{\infty}\frac{a^r}{(1-q)(1-q^2)\dots(1-q^r)}\prod_{s=0}^{\infty}(1+q^{r+s+1}).$$

SECTION 14–2

1. Setting $z=-1$ in Theorem 13–8, we see that

$$\sum_{n=-\infty}^{\infty}(-1)^n q^{n^2}=\prod_{n=0}^{\infty}(1-q^{2n+2})(1-q^{2n+1})(1-q^{2n+1})$$

$$=\prod_{n=0}^{\infty}(1-q^{n+1})\prod_{m=0}^{\infty}(1-q^{2m+1})=\prod_{n=0}^{\infty}(1-q^{n+1})\prod_{m=1}^{\infty}\frac{1}{(1+q^m)}$$

$$=\prod_{n=1}^{\infty}\frac{(1-q^n)}{(1+q^n)}.$$

INDEX OF SYMBOLS

INDEX

A CATALOG OF SELECTED

DOVER BOOKS

IN SCIENCE AND MATHEMATICS

Astronomy

CHARIOTS FOR APOLLO: The NASA History of Manned Lunar Spacecraft to 1969, Courtney G. Brooks, James M. Grimwood, and Loyd S. Swenson, Jr. This illustrated history by a trio of experts is the definitive reference on the Apollo spacecraft and lunar modules. It traces the vehicles' design, development, and operation in space. More than 100 photographs and illustrations. 576pp. 6 3/4 x 9 1/4. 0-486-46756-2

EXPLORING THE MOON THROUGH BINOCULARS AND SMALL TELESCOPES, Ernest H. Cherrington, Jr. Informative, profusely illustrated guide to locating and identifying craters, rills, seas, mountains, other lunar features. Newly revised and updated with special section of new photos. Over 100 photos and diagrams. 240pp. 8 1/4 x 11. 0-486-24491-1

WHERE NO MAN HAS GONE BEFORE: A History of NASA's Apollo Lunar Expeditions, William David Compton. Introduction by Paul Dickson. This official NASA history traces behind-the-scenes conflicts and cooperation between scientists and engineers. The first half concerns preparations for the Moon landings, and the second half documents the flights that followed Apollo 11. 1989 edition. 432pp. 7 x 10. 0-486-47888-2

APOLLO EXPEDITIONS TO THE MOON: The NASA History, Edited by Edgar M. Cortright. Official NASA publication marks the 40th anniversary of the first lunar landing and features essays by project participants recalling engineering and administrative challenges. Accessible, jargon-free accounts, highlighted by numerous illustrations. 336pp. 8 3/8 x 10 7/8. 0-486-47175-6

ON MARS: Exploration of the Red Planet, 1958-1978--The NASA History, Edward Clinton Ezell and Linda Neuman Ezell. NASA's official history chronicles the start of our explorations of our planetary neighbor. It recounts cooperation among government, industry, and academia, and it features dozens of photos from Viking cameras. 560pp. 6 3/4 x 9 1/4. 0-486-46757-0

ARISTARCHUS OF SAMOS: The Ancient Copernicus, Sir Thomas Heath. Heath's history of astronomy ranges from Homer and Hesiod to Aristarchus and includes quotes from numerous thinkers, compilers, and scholasticists from Thales and Anaximander through Pythagoras, Plato, Aristotle, and Heraclides. 34 figures. 448pp. 5 3/8 x 8 1/2. 0-486-43886-4

AN INTRODUCTION TO CELESTIAL MECHANICS, Forest Ray Moulton. Classic text still unsurpassed in presentation of fundamental principles. Covers rectilinear motion, central forces, problems of two and three bodies, much more. Includes over 200 problems, some with answers. 437pp. 5 3/8 x 8 1/2. 0-486-64687-4

BEYOND THE ATMOSPHERE: Early Years of Space Science, Homer E. Newell. This exciting survey is the work of a top NASA administrator who chronicles technological advances, the relationship of space science to general science, and the space program's social, political, and economic contexts. 528pp. 6 3/4 x 9 1/4. 0-486-47464-X

STAR LORE: Myths, Legends, and Facts, William Tyler Olcott. Captivating retellings of the origins and histories of ancient star groups include Pegasus, Ursa Major, Pleiades, signs of the zodiac, and other constellations. "Classic." – *Sky & Telescope.* 58 illustrations. 544pp. 5 3/8 x 8 1/2. 0-486-43581-4

A COMPLETE MANUAL OF AMATEUR ASTRONOMY: Tools and Techniques for Astronomical Observations, P. Clay Sherrod with Thomas L. Koed. Concise, highly readable book discusses the selection, set-up, and maintenance of a telescope; amateur studies of the sun; lunar topography and occultations; and more. 124 figures. 26 halftones. 37 tables. 335pp. 6 1/2 x 9 1/4. 0-486-42820-6

Chemistry

MOLECULAR COLLISION THEORY, M. S. Child. This high-level monograph offers an analytical treatment of classical scattering by a central force, quantum scattering by a central force, elastic scattering phase shifts, and semi-classical elastic scattering. 1974 edition. 310pp. 5 3/8 x 8 1/2. 0-486-69437-2

HANDBOOK OF COMPUTATIONAL QUANTUM CHEMISTRY, David B. Cook. This comprehensive text provides upper-level undergraduates and graduate students with an accessible introduction to the implementation of quantum ideas in molecular modeling, exploring practical applications alongside theoretical explanations. 1998 edition. 832pp. 5 3/8 x 8 1/2. 0-486-44307-8

RADIOACTIVE SUBSTANCES, Marie Curie. The celebrated scientist's thesis, which directly preceded her 1903 Nobel Prize, discusses establishing atomic character of radioactivity; extraction from pitchblende of polonium and radium; isolation of pure radium chloride; more. 96pp. 5 3/8 x 8 1/2. 0-486-42550-9

CHEMICAL MAGIC, Leonard A. Ford. Classic guide provides intriguing entertainment while elucidating sound scientific principles, with more than 100 unusual stunts: cold fire, dust explosions, a nylon rope trick, a disappearing beaker, much more. 128pp. 5 3/8 x 8 1/2. 0-486-67628-5

ALCHEMY, E. J. Holmyard. Classic study by noted authority covers 2,000 years of alchemical history: religious, mystical overtones; apparatus; signs, symbols, and secret terms; advent of scientific method, much more. Illustrated. 320pp. 5 3/8 x 8 1/2. 0-486-26298-7

CHEMICAL KINETICS AND REACTION DYNAMICS, Paul L. Houston. This text teaches the principles underlying modern chemical kinetics in a clear, direct fashion, using several examples to enhance basic understanding. Solutions to selected problems. 2001 edition. 352pp. 8 3/8 x 11. 0-486-45334-0

PROBLEMS AND SOLUTIONS IN QUANTUM CHEMISTRY AND PHYSICS, Charles S. Johnson and Lee G. Pedersen. Unusually varied problems, with detailed solutions, cover of quantum mechanics, wave mechanics, angular momentum, molecular spectroscopy, scattering theory, more. 280 problems, plus 139 supplementary exercises. 430pp. 6 1/2 x 9 1/4. 0-486-65236-X

ELEMENTS OF CHEMISTRY, Antoine Lavoisier. Monumental classic by the founder of modern chemistry features first explicit statement of law of conservation of matter in chemical change, and more. Facsimile reprint of original (1790) Kerr translation. 539pp. 5 3/8 x 8 1/2. 0-486-64624-6

MAGNETISM AND TRANSITION METAL COMPLEXES, F. E. Mabbs and D. J. Machin. A detailed view of the calculation methods involved in the magnetic properties of transition metal complexes, this volume offers sufficient background for original work in the field. 1973 edition. 240pp. 5 3/8 x 8 1/2. 0-486-46284-6

GENERAL CHEMISTRY, Linus Pauling. Revised third edition of classic first-year text by Nobel laureate. Atomic and molecular structure, quantum mechanics, statistical mechanics, thermodynamics correlated with descriptive chemistry. Problems. 992pp. 5 3/8 x 8 1/2. 0-486-65622-5

ELECTROLYTE SOLUTIONS: Second Revised Edition, R. A. Robinson and R. H. Stokes. Classic text deals primarily with measurement, interpretation of conductance, chemical potential, and diffusion in electrolyte solutions. Detailed theoretical interpretations, plus extensive tables of thermodynamic and transport properties. 1970 edition. 590pp. 5 3/8 x 8 1/2. 0-486-42225-9

Engineering

FUNDAMENTALS OF ASTRODYNAMICS, Roger R. Bate, Donald D. Mueller, and Jerry E. White. Teaching text developed by U.S. Air Force Academy develops the basic two-body and n-body equations of motion; orbit determination; classical orbital elements, coordinate transformations; differential correction; more. 1971 edition. 455pp. 5 3/8 x 8 1/2. 0-486-60061-0

INTRODUCTION TO CONTINUUM MECHANICS FOR ENGINEERS: Revised Edition, Ray M. Bowen. This self-contained text introduces classical continuum models within a modern framework. Its numerous exercises illustrate the governing principles, linearizations, and other approximations that constitute classical continuum models. 2007 edition. 320pp. 6 1/8 x 9 1/4. 0-486-47460-7

ENGINEERING MECHANICS FOR STRUCTURES, Louis L. Bucciarelli. This text explores the mechanics of solids and statics as well as the strength of materials and elasticity theory. Its many design exercises encourage creative initiative and systems thinking. 2009 edition. 320pp. 6 1/8 x 9 1/4. 0-486-46855-0

FEEDBACK CONTROL THEORY, John C. Doyle, Bruce A. Francis and Allen R. Tannenbaum. This excellent introduction to feedback control system design offers a theoretical approach that captures the essential issues and can be applied to a wide range of practical problems. 1992 edition. 224pp. 6 1/2 x 9 1/4. 0-486-46933-6

THE FORCES OF MATTER, Michael Faraday. These lectures by a famous inventor offer an easy-to-understand introduction to the interactions of the universe's physical forces. Six essays explore gravitation, cohesion, chemical affinity, heat, magnetism, and electricity. 1993 edition. 96pp. 5 3/8 x 8 1/2. 0-486-47482-8

DYNAMICS, Lawrence E. Goodman and William H. Warner. Beginning engineering text introduces calculus of vectors, particle motion, dynamics of particle systems and plane rigid bodies, technical applications in plane motions, and more. Exercises and answers in every chapter. 619pp. 5 3/8 x 8 1/2. 0-486-42006-X

ADAPTIVE FILTERING PREDICTION AND CONTROL, Graham C. Goodwin and Kwai Sang Sin. This unified survey focuses on linear discrete-time systems and explores natural extensions to nonlinear systems. It emphasizes discrete-time systems, summarizing theoretical and practical aspects of a large class of adaptive algorithms. 1984 edition. 560pp. 6 1/2 x 9 1/4. 0-486-46932-8

INDUCTANCE CALCULATIONS, Frederick W. Grover. This authoritative reference enables the design of virtually every type of inductor. It features a single simple formula for each type of inductor, together with tables containing essential numerical factors. 1946 edition. 304pp. 5 3/8 x 8 1/2. 0-486-47440-2

THERMODYNAMICS: Foundations and Applications, Elias P. Gyftopoulos and Gian Paolo Beretta. Designed by two MIT professors, this authoritative text discusses basic concepts and applications in detail, emphasizing generality, definitions, and logical consistency. More than 300 solved problems cover realistic energy systems and processes. 800pp. 6 1/8 x 9 1/4. 0-486-43932-1

THE FINITE ELEMENT METHOD: Linear Static and Dynamic Finite Element Analysis, Thomas J. R. Hughes. Text for students without in-depth mathematical training, this text includes a comprehensive presentation and analysis of algorithms of time-dependent phenomena plus beam, plate, and shell theories. Solution guide available upon request. 672pp. 6 1/2 x 9 1/4. 0-486-41181-8

HELICOPTER THEORY, Wayne Johnson. Monumental engineering text covers vertical flight, forward flight, performance, mathematics of rotating systems, rotary wing dynamics and aerodynamics, aeroelasticity, stability and control, stall, noise, and more. 189 illustrations. 1980 edition. 1089pp. 5 5/8 x 8 1/4. 0-486-68230-7

MATHEMATICAL HANDBOOK FOR SCIENTISTS AND ENGINEERS: Definitions, Theorems, and Formulas for Reference and Review, Granino A. Korn and Theresa M. Korn. Convenient access to information from every area of mathematics: Fourier transforms, Z transforms, linear and nonlinear programming, calculus of variations, random-process theory, special functions, combinatorial analysis, game theory, much more. 1152pp. 5 3/8 x 8 1/2. 0-486-41147-8

A HEAT TRANSFER TEXTBOOK: Fourth Edition, John H. Lienhard V and John H. Lienhard IV. This introduction to heat and mass transfer for engineering students features worked examples and end-of-chapter exercises. Worked examples and end-of-chapter exercises appear throughout the book, along with well-drawn, illuminating figures. 768pp. 7 x 9 1/4. 0-486-47931-5

BASIC ELECTRICITY, U.S. Bureau of Naval Personnel. Originally a training course; best nontechnical coverage. Topics include batteries, circuits, conductors, AC and DC, inductance and capacitance, generators, motors, transformers, amplifiers, etc. Many questions with answers. 349 illustrations. 1969 edition. 448pp. 6 1/2 x 9 1/4. 0-486-20973-3

BASIC ELECTRONICS, U.S. Bureau of Naval Personnel. Clear, well-illustrated introduction to electronic equipment covers numerous essential topics: electron tubes, semiconductors, electronic power supplies, tuned circuits, amplifiers, receivers, ranging and navigation systems, computers, antennas, more. 560 illustrations. 567pp. 6 1/2 x 9 1/4. 0-486-21076-6

BASIC WING AND AIRFOIL THEORY, Alan Pope. This self-contained treatment by a pioneer in the study of wind effects covers flow functions, airfoil construction and pressure distribution, finite and monoplane wings, and many other subjects. 1951 edition. 320pp. 5 3/8 x 8 1/2. 0-486-47188-8

SYNTHETIC FUELS, Ronald F. Probstein and R. Edwin Hicks. This unified presentation examines the methods and processes for converting coal, oil, shale, tar sands, and various forms of biomass into liquid, gaseous, and clean solid fuels. 1982 edition. 512pp. 6 1/8 x 9 1/4. 0-486-44977-7

THEORY OF ELASTIC STABILITY, Stephen P. Timoshenko and James M. Gere. Written by world-renowned authorities on mechanics, this classic ranges from theoretical explanations of 2- and 3-D stress and strain to practical applications such as torsion, bending, and thermal stress. 1961 edition. 560pp. 5 3/8 x 8 1/2. 0-486-47207-8

PRINCIPLES OF DIGITAL COMMUNICATION AND CODING, Andrew J. Viterbi and Jim K. Omura. This classic by two digital communications experts is geared toward students of communications theory and to designers of channels, links, terminals, modems, or networks used to transmit and receive digital messages. 1979 edition. 576pp. 6 1/8 x 9 1/4. 0-486-46901-8

LINEAR SYSTEM THEORY: The State Space Approach, Lotfi A. Zadeh and Charles A. Desoer. Written by two pioneers in the field, this exploration of the state space approach focuses on problems of stability and control, plus connections between this approach and classical techniques. 1963 edition. 656pp. 6 1/8 x 9 1/4. 0-486-46663-9

Mathematics-Bestsellers

HANDBOOK OF MATHEMATICAL FUNCTIONS: with Formulas, Graphs, and Mathematical Tables, Edited by Milton Abramowitz and Irene A. Stegun. A classic resource for working with special functions, standard trig, and exponential logarithmic definitions and extensions, it features 29 sets of tables, some to as high as 20 places. 1046pp. 8 x 10 1/2. 0-486-61272-4

ABSTRACT AND CONCRETE CATEGORIES: The Joy of Cats, Jiri Adamek, Horst Herrlich, and George E. Strecker. This up-to-date introductory treatment employs category theory to explore the theory of structures. Its unique approach stresses concrete categories and presents a systematic view of factorization structures. Numerous examples. 1990 edition, updated 2004. 528pp. 6 1/8 x 9 1/4. 0-486-46934-4

MATHEMATICS: Its Content, Methods and Meaning, A. D. Aleksandrov, A. N. Kolmogorov, and M. A. Lavrent'ev. Major survey offers comprehensive, coherent discussions of analytic geometry, algebra, differential equations, calculus of variations, functions of a complex variable, prime numbers, linear and non-Euclidean geometry, topology, functional analysis, more. 1963 edition. 1120pp. 5 3/8 x 8 1/2. 0-486-40916-3

INTRODUCTION TO VECTORS AND TENSORS: Second Edition--Two Volumes Bound as One, Ray M. Bowen and C.-C. Wang. Convenient single-volume compilation of two texts offers both introduction and in-depth survey. Geared toward engineering and science students rather than mathematicians, it focuses on physics and engineering applications. 1976 edition. 560pp. 6 1/2 x 9 1/4. 0-486-46914-X

AN INTRODUCTION TO ORTHOGONAL POLYNOMIALS, Theodore S. Chihara. Concise introduction covers general elementary theory, including the representation theorem and distribution functions, continued fractions and chain sequences, the recurrence formula, special functions, and some specific systems. 1978 edition. 272pp. 5 3/8 x 8 1/2. 0-486-47929-3

ADVANCED MATHEMATICS FOR ENGINEERS AND SCIENTISTS, Paul DuChateau. This primary text and supplemental reference focuses on linear algebra, calculus, and ordinary differential equations. Additional topics include partial differential equations and approximation methods. Includes solved problems. 1992 edition. 400pp. 7 1/2 x 9 1/4. 0-486-47930-7

PARTIAL DIFFERENTIAL EQUATIONS FOR SCIENTISTS AND ENGINEERS, Stanley J. Farlow. Practical text shows how to formulate and solve partial differential equations. Coverage of diffusion-type problems, hyperbolic-type problems, elliptic-type problems, numerical and approximate methods. Solution guide available upon request. 1982 edition. 414pp. 6 1/8 x 9 1/4. 0-486-67620-X

VARIATIONAL PRINCIPLES AND FREE-BOUNDARY PROBLEMS, Avner Friedman. Advanced graduate-level text examines variational methods in partial differential equations and illustrates their applications to free-boundary problems. Features detailed statements of standard theory of elliptic and parabolic operators. 1982 edition. 720pp. 6 1/8 x 9 1/4. 0-486-47853-X

LINEAR ANALYSIS AND REPRESENTATION THEORY, Steven A. Gaal. Unified treatment covers topics from the theory of operators and operator algebras on Hilbert spaces; integration and representation theory for topological groups; and the theory of Lie algebras, Lie groups, and transform groups. 1973 edition. 704pp. 6 1/8 x 9 1/4. 0-486-47851-3

A SURVEY OF INDUSTRIAL MATHEMATICS, Charles R. MacCluer. Students learn how to solve problems they'll encounter in their professional lives with this concise single-volume treatment. It employs MATLAB and other strategies to explore typical industrial problems. 2000 edition. 384pp. 5 3/8 x 8 1/2. 0-486-47702-9

NUMBER SYSTEMS AND THE FOUNDATIONS OF ANALYSIS, Elliott Mendelson. Geared toward undergraduate and beginning graduate students, this study explores natural numbers, integers, rational numbers, real numbers, and complex numbers. Numerous exercises and appendixes supplement the text. 1973 edition. 368pp. 5 3/8 x 8 1/2. 0-486-45792-3

A FIRST LOOK AT NUMERICAL FUNCTIONAL ANALYSIS, W. W. Sawyer. Text by renowned educator shows how problems in numerical analysis lead to concepts of functional analysis. Topics include Banach and Hilbert spaces, contraction mappings, convergence, differentiation and integration, and Euclidean space. 1978 edition. 208pp. 5 3/8 x 8 1/2. 0-486-47882-3

FRACTALS, CHAOS, POWER LAWS: Minutes from an Infinite Paradise, Manfred Schroeder. A fascinating exploration of the connections between chaos theory, physics, biology, and mathematics, this book abounds in award-winning computer graphics, optical illusions, and games that clarify memorable insights into self-similarity. 1992 edition. 448pp. 6 1/8 x 9 1/4. 0-486-47204-3

SET THEORY AND THE CONTINUUM PROBLEM, Raymond M. Smullyan and Melvin Fitting. A lucid, elegant, and complete survey of set theory, this three-part treatment explores axiomatic set theory, the consistency of the continuum hypothesis, and forcing and independence results. 1996 edition. 336pp. 6 x 9. 0-486-47484-4

DYNAMICAL SYSTEMS, Shlomo Sternberg. A pioneer in the field of dynamical systems discusses one-dimensional dynamics, differential equations, random walks, iterated function systems, symbolic dynamics, and Markov chains. Supplementary materials include PowerPoint slides and MATLAB exercises. 2010 edition. 272pp. 6 1/8 x 9 1/4. 0-486-47705-3

ORDINARY DIFFERENTIAL EQUATIONS, Morris Tenenbaum and Harry Pollard. Skillfully organized introductory text examines origin of differential equations, then defines basic terms and outlines general solution of a differential equation. Explores integrating factors; dilution and accretion problems; Laplace Transforms; Newton's Interpolation Formulas, more. 818pp. 5 3/8 x 8 1/2. 0-486-64940-7

MATROID THEORY, D. J. A. Welsh. Text by a noted expert describes standard examples and investigation results, using elementary proofs to develop basic matroid properties before advancing to a more sophisticated treatment. Includes numerous exercises. 1976 edition. 448pp. 5 3/8 x 8 1/2. 0-486-47439-9

THE CONCEPT OF A RIEMANN SURFACE, Hermann Weyl. This classic on the general history of functions combines function theory and geometry, forming the basis of the modern approach to analysis, geometry, and topology. 1955 edition. 208pp. 5 3/8 x 8 1/2. 0-486-47004-0

THE LAPLACE TRANSFORM, David Vernon Widder. This volume focuses on the Laplace and Stieltjes transforms, offering a highly theoretical treatment. Topics include fundamental formulas, the moment problem, monotonic functions, and Tauberian theorems. 1941 edition. 416pp. 5 3/8 x 8 1/2. 0-486-47755-X

Mathematics–Logic and Problem Solving

PERPLEXING PUZZLES AND TANTALIZING TEASERS, Martin Gardner. Ninety-three riddles, mazes, illusions, tricky questions, word and picture puzzles, and other challenges offer hours of entertainment for youngsters. Filled with rib-tickling drawings. Solutions. 224pp. 5 3/8 x 8 1/2. 0-486-25637-5

MY BEST MATHEMATICAL AND LOGIC PUZZLES, Martin Gardner. The noted expert selects 70 of his favorite "short" puzzles. Includes The Returning Explorer, The Mutilated Chessboard, Scrambled Box Tops, and dozens more. Complete solutions included. 96pp. 5 3/8 x 8 1/2. 0-486-28152-3

THE LADY OR THE TIGER?: and Other Logic Puzzles, Raymond M. Smullyan. Created by a renowned puzzle master, these whimsically themed challenges involve paradoxes about probability, time, and change; metapuzzles; and self-referentiality. Nineteen chapters advance in difficulty from relatively simple to highly complex. 1982 edition. 240pp. 5 3/8 x 8 1/2. 0-486-47027-X

SATAN, CANTOR AND INFINITY: Mind-Boggling Puzzles, Raymond M. Smullyan. A renowned mathematician tells stories of knights and knaves in an entertaining look at the logical precepts behind infinity, probability, time, and change. Requires a strong background in mathematics. Complete solutions. 288pp. 5 3/8 x 8 1/2. 0-486-47036-9

THE RED BOOK OF MATHEMATICAL PROBLEMS, Kenneth S. Williams and Kenneth Hardy. Handy compilation of 100 practice problems, hints and solutions indispensable for students preparing for the William Lowell Putnam and other mathematical competitions. Preface to the First Edition. Sources. 1988 edition. 192pp. 5 3/8 x 8 1/2. 0-486-69415-1

KING ARTHUR IN SEARCH OF HIS DOG AND OTHER CURIOUS PUZZLES, Raymond M. Smullyan. This fanciful, original collection for readers of all ages features arithmetic puzzles, logic problems related to crime detection, and logic and arithmetic puzzles involving King Arthur and his Dogs of the Round Table. 160pp. 5 3/8 x 8 1/2. 0-486-47435-6

UNDECIDABLE THEORIES: Studies in Logic and the Foundation of Mathematics, Alfred Tarski in collaboration with Andrzej Mostowski and Raphael M. Robinson. This well-known book by the famed logician consists of three treatises: "A General Method in Proofs of Undecidability," "Undecidability and Essential Undecidability in Mathematics," and "Undecidability of the Elementary Theory of Groups." 1953 edition. 112pp. 5 3/8 x 8 1/2. 0-486-47703-7

LOGIC FOR MATHEMATICIANS, J. Barkley Rosser. Examination of essential topics and theorems assumes no background in logic. "Undoubtedly a major addition to the literature of mathematical logic." – *Bulletin of the American Mathematical Society.* 1978 edition. 592pp. 6 1/8 x 9 1/4. 0-486-46898-4

INTRODUCTION TO PROOF IN ABSTRACT MATHEMATICS, Andrew Wohlgemuth. This undergraduate text teaches students what constitutes an acceptable proof, and it develops their ability to do proofs of routine problems as well as those requiring creative insights. 1990 edition. 384pp. 6 1/2 x 9 1/4. 0-486-47854-8

FIRST COURSE IN MATHEMATICAL LOGIC, Patrick Suppes and Shirley Hill. Rigorous introduction is simple enough in presentation and context for wide range of students. Symbolizing sentences; logical inference; truth and validity; truth tables; terms, predicates, universal quantifiers; universal specification and laws of identity; more. 288pp. 5 3/8 x 8 1/2. 0-486-42259-3

Mathematics–Algebra and Calculus

VECTOR CALCULUS, Peter Baxandall and Hans Liebeck. This introductory text offers a rigorous, comprehensive treatment. Classical theorems of vector calculus are amply illustrated with figures, worked examples, physical applications, and exercises with hints and answers. 1986 edition. 560pp. 5 3/8 x 8 1/2. 0-486-46620-5

ADVANCED CALCULUS: An Introduction to Classical Analysis, Louis Brand. A course in analysis that focuses on the functions of a real variable, this text introduces the basic concepts in their simplest setting and illustrates its teachings with numerous examples, theorems, and proofs. 1955 edition. 592pp. 5 3/8 x 8 1/2. 0-486-44548-8

ADVANCED CALCULUS, Avner Friedman. Intended for students who have already completed a one-year course in elementary calculus, this two-part treatment advances from functions of one variable to those of several variables. Solutions. 1971 edition. 432pp. 5 3/8 x 8 1/2. 0-486-45795-8

METHODS OF MATHEMATICS APPLIED TO CALCULUS, PROBABILITY, AND STATISTICS, Richard W. Hamming. This 4-part treatment begins with algebra and analytic geometry and proceeds to an exploration of the calculus of algebraic functions and transcendental functions and applications. 1985 edition. Includes 310 figures and 18 tables. 880pp. 6 1/2 x 9 1/4. 0-486-43945-3

BASIC ALGEBRA I: Second Edition, Nathan Jacobson. A classic text and standard reference for a generation, this volume covers all undergraduate algebra topics, including groups, rings, modules, Galois theory, polynomials, linear algebra, and associative algebra. 1985 edition. 528pp. 6 1/8 x 9 1/4. 0-486-47189-6

BASIC ALGEBRA II: Second Edition, Nathan Jacobson. This classic text and standard reference comprises all subjects of a first-year graduate-level course, including in-depth coverage of groups and polynomials and extensive use of categories and functors. 1989 edition. 704pp. 6 1/8 x 9 1/4. 0-486-47187-X

CALCULUS: An Intuitive and Physical Approach (Second Edition), Morris Kline. Application-oriented introduction relates the subject as closely as possible to science with explorations of the derivative; differentiation and integration of the powers of x; theorems on differentiation, antidifferentiation; the chain rule; trigonometric functions; more. Examples. 1967 edition. 960pp. 6 1/2 x 9 1/4. 0-486-40453-6

ABSTRACT ALGEBRA AND SOLUTION BY RADICALS, John E. Maxfield and Margaret W. Maxfield. Accessible advanced undergraduate-level text starts with groups, rings, fields, and polynomials and advances to Galois theory, radicals and roots of unity, and solution by radicals. Numerous examples, illustrations, exercises, appendixes. 1971 edition. 224pp. 6 1/8 x 9 1/4. 0-486-47723-1

AN INTRODUCTION TO THE THEORY OF LINEAR SPACES, Georgi E. Shilov. Translated by Richard A. Silverman. Introductory treatment offers a clear exposition of algebra, geometry, and analysis as parts of an integrated whole rather than separate subjects. Numerous examples illustrate many different fields, and problems include hints or answers. 1961 edition. 320pp. 5 3/8 x 8 1/2. 0-486-63070-6

LINEAR ALGEBRA, Georgi E. Shilov. Covers determinants, linear spaces, systems of linear equations, linear functions of a vector argument, coordinate transformations, the canonical form of the matrix of a linear operator, bilinear and quadratic forms, and more. 387pp. 5 3/8 x 8 1/2. 0-486-63518-X

Mathematics–Probability and Statistics

BASIC PROBABILITY THEORY, Robert B. Ash. This text emphasizes the probabilistic way of thinking, rather than measure-theoretic concepts. Geared toward advanced undergraduates and graduate students, it features solutions to some of the problems. 1970 edition. 352pp. 5 3/8 x 8 1/2. 0-486-46628-0

PRINCIPLES OF STATISTICS, M. G. Bulmer. Concise description of classical statistics, from basic dice probabilities to modern regression analysis. Equal stress on theory and applications. Moderate difficulty; only basic calculus required. Includes problems with answers. 252pp. 5 5/8 x 8 1/4. 0-486-63760-3

OUTLINE OF BASIC STATISTICS: Dictionary and Formulas, John E. Freund and Frank J. Williams. Handy guide includes a 70-page outline of essential statistical formulas covering grouped and ungrouped data, finite populations, probability, and more, plus over 1,000 clear, concise definitions of statistical terms. 1966 edition. 208pp. 5 3/8 x 8 1/2. 0-486-47769-X

GOOD THINKING: The Foundations of Probability and Its Applications, Irving J. Good. This in-depth treatment of probability theory by a famous British statistician explores Keynesian principles and surveys such topics as Bayesian rationality, corroboration, hypothesis testing, and mathematical tools for induction and simplicity. 1983 edition. 352pp. 5 3/8 x 8 1/2. 0-486-47438-0

INTRODUCTION TO PROBABILITY THEORY WITH CONTEMPORARY APPLICATIONS, Lester L. Helms. Extensive discussions and clear examples, written in plain language, expose students to the rules and methods of probability. Exercises foster problem-solving skills, and all problems feature step-by-step solutions. 1997 edition. 368pp. 6 1/2 x 9 1/4. 0-486-47418-6

CHANCE, LUCK, AND STATISTICS, Horace C. Levinson. In simple, non-technical language, this volume explores the fundamentals governing chance and applies them to sports, government, and business. "Clear and lively ... remarkably accurate." – *Scientific Monthly*. 384pp. 5 3/8 x 8 1/2. 0-486-41997-5

FIFTY CHALLENGING PROBLEMS IN PROBABILITY WITH SOLUTIONS, Frederick Mosteller. Remarkable puzzlers, graded in difficulty, illustrate elementary and advanced aspects of probability. These problems were selected for originality, general interest, or because they demonstrate valuable techniques. Also includes detailed solutions. 88pp. 5 3/8 x 8 1/2. 0-486-65355-2

EXPERIMENTAL STATISTICS, Mary Gibbons Natrella. A handbook for those seeking engineering information and quantitative data for designing, developing, constructing, and testing equipment. Covers the planning of experiments, the analyzing of extreme-value data; and more. 1966 edition. Index. Includes 52 figures and 76 tables. 560pp. 8 3/8 x 11. 0-486-43937-2

STOCHASTIC MODELING: Analysis and Simulation, Barry L. Nelson. Coherent introduction to techniques also offers a guide to the mathematical, numerical, and simulation tools of systems analysis. Includes formulation of models, analysis, and interpretation of results. 1995 edition. 336pp. 6 1/8 x 9 1/4. 0-486-47770-3

INTRODUCTION TO BIOSTATISTICS: Second Edition, Robert R. Sokal and F. James Rohlf. Suitable for undergraduates with a minimal background in mathematics, this introduction ranges from descriptive statistics to fundamental distributions and the testing of hypotheses. Includes numerous worked-out problems and examples. 1987 edition. 384pp. 6 1/8 x 9 1/4. 0-486-46961-1

Mathematics–Geometry and Topology

PROBLEMS AND SOLUTIONS IN EUCLIDEAN GEOMETRY, M. N. Aref and William Wernick. Based on classical principles, this book is intended for a second course in Euclidean geometry and can be used as a refresher. More than 200 problems include hints and solutions. 1968 edition. 272pp. 5 3/8 x 8 1/2. 0-486-47720-7

TOPOLOGY OF 3-MANIFOLDS AND RELATED TOPICS, Edited by M. K. Fort, Jr. With a New Introduction by Daniel Silver. Summaries and full reports from a 1961 conference discuss decompositions and subsets of 3-space; n-manifolds; knot theory; the Poincaré conjecture; and periodic maps and isotopies. Familiarity with algebraic topology required. 1962 edition. 272pp. 6 1/8 x 9 1/4. 0-486-47753-3

POINT SET TOPOLOGY, Steven A. Gaal. Suitable for a complete course in topology, this text also functions as a self-contained treatment for independent study. Additional enrichment materials make it equally valuable as a reference. 1964 edition. 336pp. 5 3/8 x 8 1/2. 0-486-47222-1

INVITATION TO GEOMETRY, Z. A. Melzak. Intended for students of many different backgrounds with only a modest knowledge of mathematics, this text features self-contained chapters that can be adapted to several types of geometry courses. 1983 edition. 240pp. 5 3/8 x 8 1/2. 0-486-46626-4

TOPOLOGY AND GEOMETRY FOR PHYSICISTS, Charles Nash and Siddhartha Sen. Written by physicists for physics students, this text assumes no detailed background in topology or geometry. Topics include differential forms, homotopy, homology, cohomology, fiber bundles, connection and covariant derivatives, and Morse theory. 1983 edition. 320pp. 5 3/8 x 8 1/2. 0-486-47852-1

BEYOND GEOMETRY: Classic Papers from Riemann to Einstein, Edited with an Introduction and Notes by Peter Pesic. This is the only English-language collection of these 8 accessible essays. They trace seminal ideas about the foundations of geometry that led to Einstein's general theory of relativity. 224pp. 6 1/8 x 9 1/4. 0-486-45350-2

GEOMETRY FROM EUCLID TO KNOTS, Saul Stahl. This text provides a historical perspective on plane geometry and covers non-neutral Euclidean geometry, circles and regular polygons, projective geometry, symmetries, inversions, informal topology, and more. Includes 1,000 practice problems. Solutions available. 2003 edition. 480pp. 6 1/8 x 9 1/4. 0-486-47459-3

TOPOLOGICAL VECTOR SPACES, DISTRIBUTIONS AND KERNELS, François Trèves. Extending beyond the boundaries of Hilbert and Banach space theory, this text focuses on key aspects of functional analysis, particularly in regard to solving partial differential equations. 1967 edition. 592pp. 5 3/8 x 8 1/2. 0-486-45352-9

INTRODUCTION TO PROJECTIVE GEOMETRY, C. R. Wylie, Jr. This introductory volume offers strong reinforcement for its teachings, with detailed examples and numerous theorems, proofs, and exercises, plus complete answers to all odd-numbered end-of-chapter problems. 1970 edition. 576pp. 6 1/8 x 9 1/4. 0-486-46895-X

FOUNDATIONS OF GEOMETRY, C. R. Wylie, Jr. Geared toward students preparing to teach high school mathematics, this text explores the principles of Euclidean and non-Euclidean geometry and covers both generalities and specifics of the axiomatic method. 1964 edition. 352pp. 6 x 9. 0-486-47214-0

Mathematics-History

THE WORKS OF ARCHIMEDES, Archimedes. Translated by Sir Thomas Heath. Complete works of ancient geometer feature such topics as the famous problems of the ratio of the areas of a cylinder and an inscribed sphere; the properties of conoids, spheroids, and spirals; more. 326pp. 5 3/8 x 8 1/2. 0-486-42084-1

THE HISTORICAL ROOTS OF ELEMENTARY MATHEMATICS, Lucas N. H. Bunt, Phillip S. Jones, and Jack D. Bedient. Exciting, hands-on approach to understanding fundamental underpinnings of modern arithmetic, algebra, geometry and number systems examines their origins in early Egyptian, Babylonian, and Greek sources. 336pp. 5 3/8 x 8 1/2. 0-486-25563-8

THE THIRTEEN BOOKS OF EUCLID'S ELEMENTS, Euclid. Contains complete English text of all 13 books of the Elements plus critical apparatus analyzing each definition, postulate, and proposition in great detail. Covers textual and linguistic matters; mathematical analyses of Euclid's ideas; classical, medieval, Renaissance and modern commentators; refutations, supports, extrapolations, reinterpretations and historical notes. 995 figures. Total of 1,425pp. All books 5 3/8 x 8 1/2.

Vol. I: 443pp. 0-486-60088-2
Vol. II: 464pp. 0-486-60089-0
Vol. III: 546pp. 0-486-60090-4

A HISTORY OF GREEK MATHEMATICS, Sir Thomas Heath. This authoritative two-volume set that covers the essentials of mathematics and features every landmark innovation and every important figure, including Euclid, Apollonius, and others. 5 3/8 x 8 1/2.

Vol. I: 461pp. 0-486-24073-8
Vol. II: 597pp. 0-486-24074-6

A MANUAL OF GREEK MATHEMATICS, Sir Thomas L. Heath. This concise but thorough history encompasses the enduring contributions of the ancient Greek mathematicians whose works form the basis of most modern mathematics. Discusses Pythagorean arithmetic, Plato, Euclid, more. 1931 edition. 576pp. 5 3/8 x 8 1/2. 0-486-43231-9

CHINESE MATHEMATICS IN THE THIRTEENTH CENTURY, Ulrich Libbrecht. An exploration of the 13th-century mathematician Ch'in, this fascinating book combines what is known of the mathematician's life with a history of his only extant work, the Shu-shu chiu-chang. 1973 edition. 592pp. 5 3/8 x 8 1/2. 0-486-44619-0

PHILOSOPHY OF MATHEMATICS AND DEDUCTIVE STRUCTURE IN EUCLID'S ELEMENTS, Ian Mueller. This text provides an understanding of the classical Greek conception of mathematics as expressed in Euclid's Elements. It focuses on philosophical, foundational, and logical questions and features helpful appendixes. 400pp. 6 1/2 x 9 1/4. 0-486-45300-6

BEYOND GEOMETRY: Classic Papers from Riemann to Einstein, Edited with an Introduction and Notes by Peter Pesic. This is the only English-language collection of these 8 accessible essays. They trace seminal ideas about the foundations of geometry that led to Einstein's general theory of relativity. 224pp. 6 1/8 x 9 1/4. 0-486-45350-2

HISTORY OF MATHEMATICS, David E. Smith. Two-volume history – from Egyptian papyri and medieval maps to modern graphs and diagrams. Non-technical chronological survey with thousands of biographical notes, critical evaluations, and contemporary opinions on over 1,100 mathematicians. 5 3/8 x 8 1/2.

Vol. I: 618pp. 0-486-20429-4
Vol. II: 736pp. 0-486-20430-8

Physics

THEORETICAL NUCLEAR PHYSICS, John M. Blatt and Victor F. Weisskopf. An uncommonly clear and cogent investigation and correlation of key aspects of theoretical nuclear physics by leading experts: the nucleus, nuclear forces, nuclear spectroscopy, two-, three- and four-body problems, nuclear reactions, beta-decay and nuclear shell structure. 896pp. 5 3/8 x 8 1/2. 0-486-66827-4

QUANTUM THEORY, David Bohm. This advanced undergraduate-level text presents the quantum theory in terms of qualitative and imaginative concepts, followed by specific applications worked out in mathematical detail. 655pp. 5 3/8 x 8 1/2. 0-486-65969-0

ATOMIC PHYSICS AND HUMAN KNOWLEDGE, Niels Bohr. Articles and speeches by the Nobel Prize–winning physicist, dating from 1934 to 1958, offer philosophical explorations of the relevance of atomic physics to many areas of human endeavor. 1961 edition. 112pp. 5 3/8 x 8 1/2. 0-486-47928-5

COSMOLOGY, Hermann Bondi. A co-developer of the steady-state theory explores his conception of the expanding universe. This historic book was among the first to present cosmology as a separate branch of physics. 1961 edition. 192pp. 5 3/8 x 8 1/2. 0-486-47483-6

LECTURES ON QUANTUM MECHANICS, Paul A. M. Dirac. Four concise, brilliant lectures on mathematical methods in quantum mechanics from Nobel Prize-winning quantum pioneer build on idea of visualizing quantum theory through the use of classical mechanics. 96pp. 5 3/8 x 8 1/2. 0-486-41713-1

THE PRINCIPLE OF RELATIVITY, Albert Einstein and Frances A. Davis. Eleven papers that forged the general and special theories of relativity include seven papers by Einstein, two by Lorentz, and one each by Minkowski and Weyl. 1923 edition. 240pp. 5 3/8 x 8 1/2. 0-486-60081-5

PHYSICS OF WAVES, William C. Elmore and Mark A. Heald. Ideal as a classroom text or for individual study, this unique one-volume overview of classical wave theory covers wave phenomena of acoustics, optics, electromagnetic radiations, and more. 477pp. 5 3/8 x 8 1/2. 0-486-64926-1

THERMODYNAMICS, Enrico Fermi. In this classic of modern science, the Nobel Laureate presents a clear treatment of systems, the First and Second Laws of Thermodynamics, entropy, thermodynamic potentials, and much more. Calculus required. 160pp. 5 3/8 x 8 1/2. 0-486-60361-X

QUANTUM THEORY OF MANY-PARTICLE SYSTEMS, Alexander L. Fetter and John Dirk Walecka. Self-contained treatment of nonrelativistic many-particle systems discusses both formalism and applications in terms of ground-state (zero-temperature) formalism, finite-temperature formalism, canonical transformations, and applications to physical systems. 1971 edition. 640pp. 5 3/8 x 8 1/2. 0-486-42827-3

QUANTUM MECHANICS AND PATH INTEGRALS: Emended Edition, Richard P. Feynman and Albert R. Hibbs. Emended by Daniel F. Styer. The Nobel Prize-winning physicist presents unique insights into his theory and its applications. Feynman starts with fundamentals and advances to the perturbation method, quantum electrodynamics, and statistical mechanics. 1965 edition, emended in 2005. 384pp. 6 1/8 x 9 1/4. 0-486-47722-3

Physics

INTRODUCTION TO MODERN OPTICS, Grant R. Fowles. A complete basic undergraduate course in modern optics for students in physics, technology, and engineering. The first half deals with classical physical optics; the second, quantum nature of light. Solutions. 336pp. 5 3/8 x 8 1/2. 0-486-65957-7

THE QUANTUM THEORY OF RADIATION: Third Edition, W. Heitler. The first comprehensive treatment of quantum physics in any language, this classic introduction to basic theory remains highly recommended and widely used, both as a text and as a reference. 1954 edition. 464pp. 5 3/8 x 8 1/2. 0-486-64558-4

QUANTUM FIELD THEORY, Claude Itzykson and Jean-Bernard Zuber. This comprehensive text begins with the standard quantization of electrodynamics and perturbative renormalization, advancing to functional methods, relativistic bound states, broken symmetries, nonabelian gauge fields, and asymptotic behavior. 1980 edition. 752pp. 6 1/2 x 9 1/4. 0-486-44568-2

FOUNDATIONS OF POTENTIAL THERY, Oliver D. Kellogg. Introduction to fundamentals of potential functions covers the force of gravity, fields of force, potentials, harmonic functions, electric images and Green's function, sequences of harmonic functions, fundamental existence theorems, and much more. 400pp. 5 3/8 x 8 1/2. 0-486-60144-7

FUNDAMENTALS OF MATHEMATICAL PHYSICS, Edgar A. Kraut. Indispensable for students of modern physics, this text provides the necessary background in mathematics to study the concepts of electromagnetic theory and quantum mechanics. 1967 edition. 480pp. 6 1/2 x 9 1/4. 0-486-45809-1

GEOMETRY AND LIGHT: The Science of Invisibility, Ulf Leonhardt and Thomas Philbin. Suitable for advanced undergraduate and graduate students of engineering, physics, and mathematics and scientific researchers of all types, this is the first authoritative text on invisibility and the science behind it. More than 100 full-color illustrations, plus exercises with solutions. 2010 edition. 288pp. 7 x 9 1/4. 0-486-47693-6

QUANTUM MECHANICS: New Approaches to Selected Topics, Harry J. Lipkin. Acclaimed as "excellent" (*Nature*) and "very original and refreshing" (*Physics Today*), these studies examine the Mössbauer effect, many-body quantum mechanics, scattering theory, Feynman diagrams, and relativistic quantum mechanics. 1973 edition. 480pp. 5 3/8 x 8 1/2. 0-486-45893-8

THEORY OF HEAT, James Clerk Maxwell. This classic sets forth the fundamentals of thermodynamics and kinetic theory simply enough to be understood by beginners, yet with enough subtlety to appeal to more advanced readers, too. 352pp. 5 3/8 x 8 1/2. 0-486-41735-2

QUANTUM MECHANICS, Albert Messiah. Subjects include formalism and its interpretation, analysis of simple systems, symmetries and invariance, methods of approximation, elements of relativistic quantum mechanics, much more. "Strongly recommended." – *American Journal of Physics*. 1152pp. 5 3/8 x 8 1/2. 0-486-40924-4

RELATIVISTIC QUANTUM FIELDS, Charles Nash. This graduate-level text contains techniques for performing calculations in quantum field theory. It focuses chiefly on the dimensional method and the renormalization group methods. Additional topics include functional integration and differentiation. 1978 edition. 240pp. 5 3/8 x 8 1/2. 0-486-47752-5

Physics

MATHEMATICAL TOOLS FOR PHYSICS, James Nearing. Encouraging students' development of intuition, this original work begins with a review of basic mathematics and advances to infinite series, complex algebra, differential equations, Fourier series, and more. 2010 edition. 496pp. 6 1/8 x 9 1/4. 0-486-48212-X

TREATISE ON THERMODYNAMICS, Max Planck. Great classic, still one of the best introductions to thermodynamics. Fundamentals, first and second principles of thermodynamics, applications to special states of equilibrium, more. Numerous worked examples. 1917 edition. 297pp. 5 3/8 x 8. 0-486-66371-X

AN INTRODUCTION TO RELATIVISTIC QUANTUM FIELD THEORY, Silvan S. Schweber. Complete, systematic, and self-contained, this text introduces modern quantum field theory. "Combines thorough knowledge with a high degree of didactic ability and a delightful style." – *Mathematical Reviews.* 1961 edition. 928pp. 5 3/8 x 8 1/2. 0-486-44228-4

THE ELECTROMAGNETIC FIELD, Albert Shadowitz. Comprehensive undergraduate text covers basics of electric and magnetic fields, building up to electromagnetic theory. Related topics include relativity theory. Over 900 problems, some with solutions. 1975 edition. 768pp. 5 5/8 x 8 1/4. 0-486-65660-8

THE PRINCIPLES OF STATISTICAL MECHANICS, Richard C. Tolman. Definitive treatise offers a concise exposition of classical statistical mechanics and a thorough elucidation of quantum statistical mechanics, plus applications of statistical mechanics to thermodynamic behavior. 1930 edition. 704pp. 5 5/8 x 8 1/4. 0-486-63896-0

INTRODUCTION TO THE PHYSICS OF FLUIDS AND SOLIDS, James S. Trefil. This interesting, informative survey by a well-known science author ranges from classical physics and geophysical topics, from the rings of Saturn and the rotation of the galaxy to underground nuclear tests. 1975 edition. 320pp. 5 3/8 x 8 1/2. 0-486-47437-2

STATISTICAL PHYSICS, Gregory H. Wannier. Classic text combines thermodynamics, statistical mechanics, and kinetic theory in one unified presentation. Topics include equilibrium statistics of special systems, kinetic theory, transport coefficients, and fluctuations. Problems with solutions. 1966 edition. 532pp. 5 3/8 x 8 1/2. 0-486-65401-X

SPACE, TIME, MATTER, Hermann Weyl. Excellent introduction probes deeply into Euclidean space, Riemann's space, Einstein's general relativity, gravitational waves and energy, and laws of conservation. "A classic of physics." – *British Journal for Philosophy and Science.* 330pp. 5 3/8 x 8 1/2. 0-486-60267-2

RANDOM VIBRATIONS: Theory and Practice, Paul H. Wirsching, Thomas L. Paez and Keith Ortiz. Comprehensive text and reference covers topics in probability, statistics, and random processes, plus methods for analyzing and controlling random vibrations. Suitable for graduate students and mechanical, structural, and aerospace engineers. 1995 edition. 464pp. 5 3/8 x 8 1/2. 0-486-45015-5

PHYSICS OF SHOCK WAVES AND HIGH-TEMPERATURE HYDRO DYNAMIC PHENOMENA, Ya B. Zel'dovich and Yu P. Raizer. Physical, chemical processes in gases at high temperatures are focus of outstanding text, which combines material from gas dynamics, shock-wave theory, thermodynamics and statistical physics, other fields. 284 illustrations. 1966–1967 edition. 944pp. 6 1/8 x 9 1/4. 0-486-42002-7